Annals of Mathematics Studies

Number 157

On the Tangent Space to the Space of Algebraic Cycles on a Smooth Algebraic Variety

Mark Green and Phillip Griffiths

PRINCETON UNIVERSITY PRESS
PRINCETON AND OXFORD

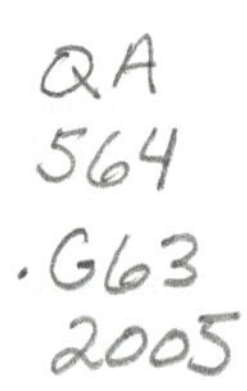

Published by Princeton University Press, 41 William Street,
Princeton, New Jersey 08540
In the United Kingdom: Princeton University Press, 3 Market Place,
Woodstock, Oxfordshire OX20 1SY

Library of Congress Control Number 2004108912

ISBN: 0-691-12043-9
0-691-12044-7 (paper)

British Library Cataloging-in-Publication Data is available

The publisher would like to acknowledge the authors of this
volume for providing the camera-ready copy from which this
book was printed

Printed on acid-free paper

pup.princeton.edu

Printed in the United States of America

10 9 8 7 6 5 4 3 2 1

Contents

Abstract

In this work we shall propose definitions for the tangent spaces $TZ^n(X)$ and $TZ^1(X)$ to the groups $Z^n(X)$ and $Z^1(X)$ of 0-cycles and divisors, respectively, on a smooth n-dimensional algebraic variety. Although the definitions are algebraic and formal, the motivation behind them is quite geometric and much of the text is devoted to this point. It is noteworthy that both the regular differential forms of all degrees and the field of definition enter significantly into the definition. An interesting and subtle algebraic point centers around the construction of the map $T\mathrm{Hilb}^p(X) \to TZ^p(X)$. Another interesting algebraic/geometric point is the necessary appearance of spreads and absolute differentials in higher codimension.

For an algebraic surface X we shall also define the subspace $TZ^2_{\mathrm{rat}}(X) \subset TZ^2(X)$ of tangents to rational equivalences, and we shall show that there is a natural isomorphism

$$T_f CH^2(X) \cong TZ^2(X)/TZ^2_{\mathrm{rat}}(X)$$

where the left-hand side is the formal tangent space to the Chow groups defined by Bloch. This result gives a geometric existence theorem, albeit at the infinitesimal level. The "integration" of the infinitesimal results raises very interesting geometric and arithmetic issues that are discussed at various places in the text.

Chapter One

Introduction

1.1 GENERAL COMMENTS

In this work we shall define the tangent spaces

$$TZ^n(X)$$

and

$$TZ^1(X)$$

to the spaces of 0-cycles and of divisors on a smooth, n-dimensional complex algebraic variety X. We think it may be possible to use similar methods to define $TZ^p(X)$ for all codimensions, but we have not been able to do this because of one significant technical point. Although the final definitions, as given in sections 7 and 8 below, are algebraic and formal, the motivation behind them is quite geometric. This is explained in the earlier sections; we have chosen to present the exposition in the monograph following the evolution of our geometric understanding of what the tangent spaces should be rather than beginning with the formal definition and then retracing the steps leading to the geometry.

Briefly, for 0-cycles an *arc* in $Z^n(X)$ is given by a $\mathbb{Z}$-linear combination of arcs in the symmetric products $X^{(d)}$, where such an arc is given by a smooth algebraic curve B together with a regular map $B \to X^{(d)}$. If t is a local uniformizing parameter on B we shall use the notation $t \to x_1(t)+\cdots+x_d(t)$ for the arc in $X^{(d)}$. Arcs in $Z^n(X)$ will be denoted by $z(t)$. We set $|z(t)| =$ support of $z(t)$, and if $0 \in B$ is a reference point we denote by $Z^n_{\{x\}}(X)$ the subgroup of arcs in $Z^n(X)$ with $\lim_{t\to 0}|z(t)| = x$. The tangent space will then be defined to be

$$TZ^n(X) = \left\{\text{arcs in } Z^n(X)\right\} / \equiv_{1^{\text{st}}},$$

where $\equiv_{1^{\text{st}}}$ is an equivalence relation. Although we think it should be possible to define $\equiv_{1^{\text{st}}}$ axiomatically, as in differential geometry, we have only been able to do this in special cases.

Among the main points uncovered in our study we mention the following:

(a) *The tangent space to the space of algebraic cycles is quite different from—and in some ways richer than—the tangent space to Hilbert schemes.*

This reflects the group structure on $Z^p(X)$ and properties such as

$$\begin{cases} (z(t)+\widetilde{z}(t))' = z'(t)+\widetilde{z}'(t), \\ (-z(t))' = -z'(t), \end{cases} \tag{1.1}$$

where $z(t)$ and $\tilde{z}(t)$ are arcs in $Z^p(X)$ with respective tangents $z'(t)$ and $\tilde{z}'(t)$. As a simple illustration, on a surface X an irreducible curve Y with a normal vector field ν may be obstructed in $\mathrm{Hilb}^1(X)$—e.g., the first order variation of Y in X given by ν may not be extendable to second order. However, considering Y in $Z^1(X)$ as a codimension-1 cycle the first order variation given by ν extends to second order. In fact, it can be shown that both $TZ^1(X)$ and $TZ^n(X)$ are smooth, in the sense that for $p = 1, n$ every map $\mathrm{Spec}(\mathbb{C}[\epsilon]/\epsilon^2) \to Z^p(X)$ is tangent to a geometric arc in $Z^p(X)$.

For the second point, it is well known that algebraic cycles in codimension $p \geqq 2$ behave quite differently from the classical case $p = 1$ of divisors. It turns out that infinitesimally this difference is reflected in a very geometric and computable fashion. In particular,

(b) *The differentials $\Omega^k_{X/\mathbb{C}}$ for all degrees k with $1 \leqq k \leqq n$ necessarily enter into the definition of $TZ^n(X)$.*

Remark that a tangent to the Hilbert scheme at a smooth point is uniquely determined by evaluating 1-forms on the corresponding normal vector field to the subscheme. However, for $Z^n(X)$ the forms of all degrees are required when we want to evaluate on a tangent vector, and it is in this sense that again the tangent space to the space of 0-cycles has a richer structure than the Hilbert scheme. Moreover, we see in (b) that the geometry of higher codimensional algebraic cycles is fundamentally different from that of divisors.

A third point is the following: For an algebraic curve one may give the definition of $TZ^1(X)$ either complex-analytically or algebro-geometrically with equivalent end results. However, it turns out that

(c) *For $n \geqq 2$, even if one is only interested in the complex geometry of X the field of definition of an arc $z(t)$ in $Z^n(X)$ necessarily enters into the description of $z'(0)$.*

Thus, although one may formally define $TZ^n(X)$ in the analytic category, it is only in the algebraic setting that the definition is satisfactory from a geometric perspective. One reason is the following: Any reasonable set of axiomatic properties on first order equivalence of arcs in $Z^n(X)$—including (1.1) above—leads for $n \geqq 2$ to the defining relations for absolute Kähler differentials (cf. section 6.2 below). However, only in the algebraic setting is it the case that the sheaf of Kähler differentials over $\mathbb{C}$ coincides with the sheaf of sections of the cotangent bundle (essentially, one cannot differentiate an infinite series term by term using Kähler differentials). For subtle geometric reasons, (b) and (c) turn out to be closely related.

(d) *A fourth significant difference between divisors and higher codimensional cycles is the following: For divisors it is the case that*

$$\textit{If } z_{u_k} \equiv_{\mathrm{rat}} 0 \textit{ for a sequence } u_k \textit{ tending to } 0, \textit{ then } z_0 \equiv_{\mathrm{rat}} 0.$$

For higher codimension this is false; rational equivalence has an intrinsic "graininess" in codimension $\geqq 2$. If one enhances rational equivalence by closing it up under this property, one obtains the kernel of the Abel-Jacobi map. As will be seen in the text, this graininess in codimension $\geqq 2$ manifests itself in the tangent spaces to cycles in that absolute differentials appear. This is related to the spread construction referred to later in this introduction.

(e) *Although creation/annihilation arcs are present for divisors on curves, they play a relatively inessential role. However, for $n \geqq 2$ it is crucial to understand the infinitesimal behavior of creation/annihilation arcs as these represent the tangencies to "irrelevant" rational equivalences which, in some sense, are the key new aspects in the study of higher codimensional cycles.*

One may of course quite reasonably ask:

Why should one **want** *to define $TZ^p(X)$?*

One reason is that we wanted to understand if there is geometric significance to Spencer Bloch's expression for the formal tangent space to the higher Chow groups, in which absolute differentials mysteriously appear. One of our main results is a response to this question, given by Theorem (8.47) in section 8.3 below. A perhaps deeper reason is the following: The basic Hodge-theoretic invariants of an algebraic cycle are expressed by integrals which are generally transcendental functions of the algebraic parameters describing the cycle. Some of the most satisfactory studies of these integrals have been when they satisfy some sort of functional equation, as is the situation for elliptic functions. However, this will not be the case in general. The other most fruitful approach has been by infinitesimal methods, such as the Picard-Fuchs differential equations and infinitesimal period relations (including the infinitesimal form of functional equations), both of which are of an algebraic character. Just as the infinitesimal period relations for variation of Hodge structure are expressed in terms of the tangent spaces to moduli, it seemed to us desirable to be able to express the infinitesimal Hodge-theoretic invariants of an algebraic cycle—especially those beyond the usual Abel-Jacobi images—in terms of the tangent spaces to cycles. In this monograph we will give such an expression for 0-cycles on a surface.

In the remainder of this introduction we shall summarize the different parts of this book and in so doing explain in more detail the above points.

In chapter 2 we begin by defining $TZ^1(X)$ when X is a smooth algebraic curve, a case that is both suggestive and misleading. Intuitively, we consider arcs $z(t)$ in the space $Z^1(X)$ of 0-cycles on X, and we want to define an equivalence relation $\equiv_{1^{st}}$ on such arcs so that

$$TZ^1(X) = \{\text{set of arcs in } Z^1(X)\}/\equiv_{1^{st}} .$$

The considerations are clearly local,[1] and locally we may take

$$z(t) = \operatorname{div} f(t)$$

where $f(t)$ is an arc in $\mathbb{C}(X)^*$. We set $|z(t)|$ = "support of $z(t)$" and assume that $\lim_{t\to 0} |z(t)| = x$. Writing $f(t) = f + tg + \cdots$ elementary geometric considerations suggest that, with the obvious notation, we should define

$$\operatorname{div} f(t) \equiv_{1^{st}} \operatorname{div} \widetilde{f}(t) \Leftrightarrow [g/f]_x = [\widetilde{g}/\widetilde{f}]_x$$

[1]Throughout this work, unless stated otherwise we will use the Zariski topology. We also denote by $\mathbb{C}(X)$ the field of rational functions on X, with $\mathbb{C}(X)^*$ being the multiplicative group of nonzero functions.

where $[h]_x$ denotes the principal part of the rational function h at the point $x \in X$. Thus, letting $T_{\{x\}}Z^1(X) := TZ^1_{\{x\}}(X)$ be the tangents to arcs $z(t)$ with $\lim_{t\to 0} |z(t)| = x$, we have as a provisional description

$$T_{\{x\}}Z^1(X) = \mathcal{PP}_{X,x}. \tag{1.2}$$

where $\mathcal{PP}_{X,x} = \underline{\underline{\mathbb{C}}}(X)_x/\mathcal{O}_{X,x}$ is the stalk at x of the sheaf of principal parts.

Another possible description of $T_{\{x\}}Z^1(X)$ is suggested by the classical theory of abelian sums. Namely, working in a neighborhood of $x \in X$ and writing

$$z(t) = \sum_i n_i x_i(t),$$

for $\omega \in \Omega^1_{X/\mathbb{C},x}$ we set

$$I(z,\omega) = \frac{d}{dt}\left(\sum_i n_i \int_x^{x_i(t)} \omega\right)_{t=0}.$$

Then $I(z,\omega)$ should depend only on the equivalence class of $z(t)$, and in fact we show that

$$I(z,\omega) = \operatorname{Res}_x(z'\omega)$$

where $z' \in \mathcal{PP}_{X,x}$ is the tangent to $z(t)$ using the description (1.2). This leads to a nondegenerate pairing

$$T_{\{x\}}Z^1(X) \otimes_{\mathbb{C}} \Omega^1_{X/\mathbb{C},x} \to \mathbb{C}$$

so that with either of the above descriptions we have

$$T_{\{x\}}Z^1(X) \cong \operatorname{Hom}^c_{\mathbb{C}}(\Omega^1_{X/\mathbb{C},x}, \mathbb{C}), \tag{1.3}$$

where $\operatorname{Hom}^c_{\mathbb{C}}(\Omega^1_{X/\mathbb{C},x}, \mathbb{C})$ are the continuous homomorphisms in the $\mathfrak{m}_x$-adic topology.

Now (1.3) suggests "duality," and indeed it is easy to see that a third possible description

$$T_{\{x\}}Z^1(X) \cong \lim_{i\to\infty} \mathcal{E}xt^1_{\mathcal{O}_{X,x}}(\mathcal{O}_X/\mathfrak{m}^i_x, \mathcal{O}_X) \tag{1.4}$$

is valid. Of the three descriptions (1.2)–(1.4) of $T_{\{x\}}Z^1(X)$, it will turn out that (1.3) and (1.4) suggest the correct extensions to the case of 0-cycles on n-dimensional varieties. However, the extension is not straightforward. For example, one might suspect that similar consideration of abelian sums would lead to the description (1.3) using 1-forms in general. For interesting geometric reasons, this turns out not to be correct, since, as was suggested above and will be explained below, the correct notion of abelian sums will involve integrals of differential forms of all degrees. Thus, on a smooth variety of dimension n the analogue of the right-hand side of (1.3) will give only part of the tangent space.

As will be explained below, (1.4) also extends but again not in the obvious way. The correct extension which gives the *formal definitions* of the tangent spaces $T_{\{x\}}Z^n(X)$ and tangent sheaf $\underline{\underline{T}}Z^n(X)$ is

$$T_{\{x\}}Z^n(X) := \lim_{i\to\infty} \mathcal{E}xt^n_{\mathcal{O}_{X,x}}(\mathcal{O}_X/\mathfrak{m}^i_x, \Omega^{n-1}_{X/\mathbb{Q}}) \tag{1.5}$$

and

$$\underline{\underline{T}}Z^n(X) = \bigoplus_{x\in X} \lim_{i\to\infty} \underline{\underline{\mathcal{E}xt}}^n_{\mathcal{O}_X}(\mathcal{O}_X/\mathfrak{m}^i_x, \Omega^{n-1}_{X/\mathbb{Q}}).$$

The geometric reasons why absolute differentials appear have to do with the points (b) and (c) above and will be discussed below.

The basic building blocks for 0-cycles on a smooth variety X are the *configuration spaces* consisting of sets of m points x_i on X. As a variety this is just the mth symmetric product $X^{(m)}$, whose points we write as effective 0-cycles

$$z = x_1 + \cdots + x_m.$$

We wish to study the geometry of the $X^{(m)}$ collectively, and for this we are interested in differential forms φ_m on $X^{(m)}$ that have the *hereditary property*

$$\varphi_{m+1}\Big|X^{(m)}_x = \varphi_m \tag{1.6}$$

where for each fixed $x \in X$ the inclusion $X^{(m)}_x \hookrightarrow X^{(m+1)}$ is given by $z \to z + x$. One such collection of differential forms on the various $X^{(m)}$'s is given by the *traces* $\operatorname{Tr}\varphi$ of a form $\varphi \in \Omega^q_{X/\mathbb{C}}$. Here, we come to the first geometric reason why forms of higher degree necessarily enter when $n \geqq 1$:

(1.7) $\Omega^*_{X^{(m)}/\mathbb{C}}$ *is generated over* $\mathcal{O}_{X^{(m)}}$ *by sums of elements of the form*

$$\operatorname{Tr}\omega_1 \wedge \cdots \wedge \operatorname{Tr}\omega_k, \qquad \omega_i \in \Omega^{q_i}_{X/\mathbb{C}}.$$

Moreover, we **must** *add generators* $\operatorname{Tr}\omega$ *where* $\omega \in \Omega^q_{X/\mathbb{C}}$ *for all* q *with* $1 \leqq q \leqq n$ *to reach all of* $\Omega^*_{X^{(m)}/\mathbb{C}}$.

Of course, forms of higher degree are needed only in neighborhoods of singular points on the $X^{(m)}$, and for $n \geqq 2$ the singular locus is exactly the diagonals where two or more points coincide.

Put differently, the structure of point configurations is reflected by the geometry of the $X^{(m)}$. The infinitesimal structure of point configurations is then reflected along the diagonals where two or more points have come together and where the $X^{(m)}$ are singular for $n \geqq 2$. The geometric properties of point configurations are in turn reflected by the regular differential forms on the symmetric products, particularly those having the hereditary property (1.6). There is *new* geometric information measured by the traces of q-forms for each q with $1 \leqq q \leqq n$, and thus *the definition of the tangent space to 0-cycles should involve the differential forms of all degrees*. This is clearly illustrated by the coordinate calculations given in chapter 3.

What is this new geometric information reflected by the differential forms of higher degree? One answer stems from E. Cartan, who taught us that when there are natural parameters in a geometric structure then those parameters should be included as part of that structure. In the present situation, if in terms of local uniformizing parameters on B and on X we represent arcs in the space of 0-cycles as sums of Puiseaux series, then the coefficients of these series provide natural parameters for the space of arcs in $Z^n(X)$. It turns out that for $n \geqq 2$ there is new infinitesimal information in these parameters arising from the higher degree forms on X. This phenomenon occurs only in higher codimension and is an essential ingredient in the geometric understanding of the infinitesimal structure of higher codimensional cycles.

The traces of forms $\omega \in \Omega^q_{X/\mathbb{C}}$ give rise to what are provisionally called universal abelian invariants $\widetilde{I}(z,\omega)$ (cf. chapter 3), which in coordinates are certain expressions in the Puiseaux coefficients and their differentials of degree $q-1$. In order to define the relation of equivalence of arcs in the space of 0-cycles what is needed is some way to map the $q-1$ forms in the Puiseaux coefficients to a fixed vector space; i.e., a method of comparing the infinitesimal structure at different arcs. Such a map exists, provided that instead of the usual differential forms we take *absolute differentials*. Recall that for any algebraic or analytic variety Y and any subfield k of the complex numbers we may define the Kähler differentials over k of degree r, denoted $\Omega^r_{\mathcal{O}_Y/k}$. For any subvariety $W \subset Y$ there are restriction maps

$$\Omega^r_{\mathcal{O}_Y/k} \to \Omega^r_{\mathcal{O}_W/k}.$$

Taking W to be a point $y \in Y$ and the field k to be $\mathbb{Q}$, since $\mathcal{O}_y \cong \mathbb{C}$ there is an evaluation map

$$e_y : \Omega^r_{\mathcal{O}_{Y,y}/\mathbb{Q}} \to \Omega^r_{\mathbb{C}/\mathbb{Q}}. \tag{1.9}$$

Applying this when Y is the space of Puiseaux coefficients, for an arc z in the space of 0-cycles and form $\omega \in \Omega^q_{\mathcal{O}_X/\mathbb{Q}}$ we may finally define the *universal abelian invariants*

$$I(z,\omega) = e_z \widetilde{I}(z,\omega).$$

Two arcs z and $\widetilde{z}$ are said to be *geometrically equivalent to first order*, written $z \equiv_{1^{\text{st}}} \widetilde{z}$, if

$$I(z,\omega) = I(\widetilde{z},\omega)$$

for all $\omega \in \Omega^q_{\mathcal{O}_X/\mathbb{Q}}$ and all q with $1 \leqq q \leqq n$. It turns out that here it is sufficient to consider only $\omega \in \Omega^n_{X/\mathbb{Q}}$. The space is filtered with $Gr^q \Omega^n_{X/\mathbb{Q}} \cong \Omega^{n-q}_{\mathbb{C}/\mathbb{Q}} \otimes \Omega^1_{X/\mathbb{C}}$, and roughly speaking we may think of $\Omega^n_{X/\mathbb{Q}}$ as encoding the information in the $\Omega^q_{X/\mathbb{C}}$'s for $1 \leqq q \leqq n$. Intuitively, $\equiv_{1^{\text{st}}}$ captures the invariant information in the differentials at $t=0$ of Puiseaux series, where the coefficients are differentiated in the sense of $\Omega^1_{\mathbb{C}/\mathbb{Q}}$. The simplest interesting case is when X is a surface defined over $\mathbb{Q}$, ξ and $\eta \in \mathbb{Q}(X)$ give local uniformizing parameters, and $z(t)$ is an arc in $Z^2(X)$ given by

$$z(t) = z_+(t) + z_-(t)$$

where

$$z_\pm(t) = (\xi_\pm(t), \eta_\pm(t))$$

with

$$\begin{cases} \xi_\pm(t) = \pm a_1 t^{1/2} + a_2 t + \cdots \\ \eta_\pm(t) = \pm b_1 t^{1/2} + b_2 t + \cdots . \end{cases}$$

The information in the universal abelian invariants $I(z, \varphi)$ for $\varphi \in \Omega^1_{X/\mathbb{C}}$ is

$$a_1^2, a_1 b_1, b_1^2; a_2, b_2.$$

The additional information in $I(z, \omega)$ for $\omega \in \Omega^2_{X/\mathbb{C}}$ is

$$a_1 db_1 - b_1 da_1,$$

which is not a consequence of the differentials of the $I(z, \varphi)$'s for $\varphi \in \Omega^1_{X/\mathbb{C}}$.

We then give a geometric description of the tangent space as

$$TZ^n(X) = \{\text{arcs in } Z^n(X)\} / \equiv_{1^{\text{st}}} .$$

The calculations given in chapter 3 show that this definition is independent of the particular coordinate system used to define the space of Puiseaux coefficients. We emphasize that this is *not* the formal definition of $TZ^n(X)$—that definition is given by (1.5), and as we shall show it is equivalent to the above geometric description.

So far this discussion applies to the analytic as well as to the algebraic category. However, *only* in the algebraic setting is it the case that

$$\Omega^1_{\mathcal{O}_X/\mathbb{C}} \cong \mathcal{O}_X(T^*X); \tag{1.10}$$

that is, only in the algebraic setting is it the case that Kähler differentials over $\mathbb{C}$ give the right geometric object. Thus, the above sleight of hand where we used Kähler differentials to define the universal abelian invariants

$$I(z, \omega) \in \Omega^{q-1}_{\mathbb{C}/\mathbb{Q}}$$

will only give the correct geometric notion in the algebraic category. In chapter 4 we give a heuristic, computational approach to absolute differentials. In particular we explain why (1.10) only works in the algebraic setting. The essential point is that the axioms for Kähler differentials extend to allow term by term differentiation of the power series expansion of an *algebraic* function, but this does not hold for a general *analytic* function.

In the algebraic setting $\Omega^1_{\mathcal{O}_X/\mathbb{Q}} = \Omega^1_{X/\mathbb{Q}}$, and there is an additional geometric interpretation of the "arithmetic part" $\Omega^1_{\mathbb{C}/\mathbb{Q}} \otimes \mathcal{O}_X$ of the absolute differentials $\Omega^1_{X/\mathbb{Q}}$. This deals with the notion of a *spread*, and again in chapter 4 we give a heuristic, geometric discussion of this concept. Given a 0-cycle z on an algebraic variety X, both defined over a field k that is finitely generated over $\mathbb{Q}$, the spread will be a family

$$\begin{array}{ccc} \mathcal{X} & \supset & \mathcal{Z} \\ \downarrow & & \downarrow \\ S & = & S \end{array}$$

$$\mathcal{X} = \{X_s\}_{s \in S}, \ \mathcal{Z} = \{z_s\}_{s \in S}$$

where $\mathcal{X}$, $\mathcal{Z}$, and S are all defined over $\mathbb{Q}$ and $\mathbb{Q}(S) \cong k$, and where the fiber over a generic point $s_0 \in S$ is our original X and z. Roughly speaking we may think of spreads as arising from the different embeddings of k into $\mathbb{C}$; thus, for $s \in S$ not lying in a proper subvariety defined over $\mathbb{Q}$ the algebraic properties of X_s and z_s are the same as those of X and z. There is a canonical mapping

$$T^*_{s_0}S \to \Omega^1_{k/\mathbb{Q}}, \tag{1.11}$$

and under this mapping the extension class of

$$0 \to \Omega^1_{k/\mathbb{Q}} \otimes_k \mathcal{O}_{X(k)} \to \Omega^1_{X(k)/\mathbb{Q}} \to \Omega^1_{X(k)/k} \to 0$$

corresponds to the Kodaira-Spencer class of the family $\{X_s\}_{s\in S}$ at s_0. The fact that the spread gives in higher codimension the natural parameters of a cycle and that infinitesimally the spread is expressed in terms of $\Omega^1_{X/\mathbb{Q}}$ are two reasons why absolute differentials necessarily appear.

Using this discussion of absolute differentials and spreads in chapter 4, in chapter 5 we turn to the geometric description of the tangent space $TZ^n(X)$ to the space of 0-cycles or a smooth n-dimensional algebraic variety X. We say "geometric" because the formal algebraic definition of the tangent spaces $TZ^n(X)$ will be given in chapter 7 using an extension of the *Ext* construction discussed above in the $n = 1$ case. This definition will then be proved to coincide with the description using the universal abelian invariants discussed above. In chapter 5 we give an alternate, intrinsic definition of the $I(z, \omega)$'s based on functorial properties of absolute differentials.

In chapter 4 we have introduced absolute differentials as a means of mapping the differentials of the parameters of an arc in $Z^n(X)$ expressed in terms of local uniformizing parameters to a reference object. Geometrically, using (1.11) this construction reflects infinitesimal variation in the spread directions. Algebraically, for an algebraic variety Y and point $y \in Y$, the evaluation mapping (1.9) is for $r = 1$ given by

$$f\,dg \xrightarrow{e_y} f(y)d(g(y)),$$

where $d = d_{\mathbb{C}/\mathbb{Q}}$ and $f, g \in \mathbb{C}(Y)$ are rational functions on Y that are regular near y. If Y is defined over $\mathbb{Q}$ and $f, g \in \mathbb{Q}(Y)$, then e_y reflects the field of definition of y—it is thus measuring arithmetic information.

One may reasonably ask: *Is there an alternative, purely geometric way of defining* $\equiv_{1^{\text{st}}}$ *for arcs in* $Z^n(X)$ *that leads to absolute differentials?* In other words, even if one is only interested in the complex geometry of the space of 0-cycles, is there a geometric reason why arithmetic considerations enter the picture? Although we have not been able to completely define $\equiv_{1^{\text{st}}}$ axiomatically, we suspect that this can be done and in a number of places we will show geometrically how differentials over $\mathbb{Q}$ necessarily arise.

For example, in section 6.2 we consider the free group F generated by the arcs in $Z^2(\mathbb{C}^2)$ given by differences $z_{\alpha\beta}(t) - z_{1\beta}(t)$ where $z_{\alpha\beta}(t)$ is the 0-cycle given by the equations

$$z_{\alpha\beta}(t) = \begin{cases} x^2 - \alpha y^2 = 0 & \alpha \neq 0 \\ xy - \beta t = 0. \end{cases}$$

There we list a set of "evident" geometric axioms for first order equivalence of arcs in F, and then an elementary but somewhat intricate calculation shows that the map

$$F/\equiv_{1^{\text{st}}} \to \Omega^1_{\mathbb{C}/\mathbb{Q}}$$

given by

$$z_{\alpha\beta}(t) \to \beta\frac{d\alpha}{\alpha} \qquad (d \text{ denotes } d_{\mathbb{C}/\mathbb{Q}})$$

is a well-defined isomorphism. Essentially, the condition that the tangent map be a homomorphism to a vector space that factors through the tangent map to the Hilbert scheme leads directly to the defining relations for absolute Kähler differentials.

Another example, one that will be used elsewhere in the book, begins with the observation that $Z^n(X)$ is the group of global sections of the Zariski sheaf

$$\bigoplus_{x\in X} \underline{\underline{\mathbb{Z}}}_x.$$

Taking X to be a curve we may consider the Zariski sheaf

$$\bigoplus_{x\in X} \underline{\underline{\mathbb{C}}}^*_x$$

whose global sections we denote by $Z_1^1(X)$. This sheaf arises naturally when one localizes the tame symbol mappings T_x that arise in the Weil reciprocity law. In section 6.2 we give a set of geometric axioms on arcs in $Z_1^1(X)$ that define an equivalence relation yielding a description of the sheaf $\underline{\underline{T}}Z_1^1(X)$ as

$$\underline{\underline{T}}Z_1^1(X) \cong \bigoplus_{x\in X} \underline{\underline{\mathrm{Hom}}}^o(\Omega^1_{X/\mathbb{Q},x}, \Omega^1_{\mathbb{C}/\mathbb{Q}});$$

here $\mathrm{Hom}^o(\Omega^1_{X/\mathbb{Q},x}, \Omega^1_{\mathbb{C}/\mathbb{Q}})$ are the continuous $\mathbb{C}$-linear homomorphims $\Omega^1_{X/\mathbb{Q},x} \xrightarrow{\varphi} \Omega^1_{\mathbb{C}/\mathbb{Q}}$ that satisfy

$$\varphi(f\alpha) = \varphi_0(f)\alpha,$$

where $f \in \mathcal{O}_{X,x}$, $\alpha \in \Omega^1_{\mathbb{C}/\mathbb{Q}}$, and $\varphi_0 : \mathcal{O}_{X,x} \to \mathbb{C}$ is a continuous $\mathbb{C}$-linear homomorphism. The point is that again purely geometric considerations lead naturally to differentials over $\mathbb{Q}$. Essentially, the reason again comes down to the assumption that

$$(z \pm \widetilde{z})' = z' \pm \widetilde{z}'$$

i.e., the tangent map should be a homomorphism from arcs in $Z_1^1(X)$ to a vector space.

In (d) at the beginning of the introduction we mentioned different limiting properties of rational equivalence for divisors and higher codimensional cycles. One property that our tangent space construction has is the following: Let $z_u(t)$ be a family of arcs in $Z^p(X)$. Then

$$\textit{If } z_u'(0) = 0 \textit{ for all } u \neq 0, \textit{ then } z_0'(0) = 0.$$

Once again the statement

$$\lim z_{u_k}'(0) = 0 \textit{ for a sequence } u_k \to 0 \textit{ implies that } z_0'(0) = 0$$

is true for divisors but false in higher codimension. The reason is essentially this: Any algebraic construction concerning algebraic cycles survives when we take the spread of the variety together with the cycles over their field of definition. Geometric invariants arising in the spread give invariants of the original cycle. Infinitesimally, related to (b) above there is in higher codimensions new information arising from evaluating q-forms ($q \geqq 2$) on multivectors $v \wedge w_1 \wedge \cdots \wedge w_{q-1}$, where v is the tangent to the arc in the usual "geometric" sense and $w_1, \ldots, w_{q-1}$ are tangents in the spread directions. Thus arithmetic considerations appear at the level of the tangent space to cycles (we did not expect this) and survive in the tangent space to Chow groups where they appear in Bloch's formula.

Above, we mentioned the tame symbol $T_x(f, g) \in \mathbb{C}^*$ of $f, g \in \mathbb{C}(X)^*$. It has the directly verified properties

$$\begin{cases} T_x(f^m, g) = T_x(f, g)^m \text{ for } m \in \mathbb{Z} \\ T_x(f_1 f_2, g) = T_x(f_1, g) T_x(f_2, g) \\ T_x(f, g) = T_x(g, f)^{-1} \\ T_x(f, 1-f) = 1, \end{cases}$$

which show that the tame symbol gives mappings

$$T_x : K_2\big(\mathbb{C}(X)\big) \to \mathbb{C}^*, \text{ and } T : K_2(\mathbb{C}(X)) \to \bigoplus_x \mathbb{C}^*_x.$$

A natural question related to the definition of $\underline{\underline{T}} Z^1_1(X)$ is

What is the differential of the tame symbol?

According to van der Kallen [12], for any field or local ring F in characteristic zero the formal tangent space to $K_2(F)$ is given by

$$TK_2(F) \cong \Omega^1_{F/\mathbb{Q}}. \tag{1.12}$$

Thus, we are seeking to calculate

$$\Omega^1_{\mathbb{C}(X)/\mathbb{Q}} \xrightarrow{dT_x} \operatorname{Hom}^o(\Omega^1_{X/\mathbb{Q},x}, \Omega^1_{\mathbb{C}/\mathbb{Q}}).$$

In section 6.3 we give this evaluation in terms of residues; this calculation again illustrates the linking of arithmetic and geometry. As an aside, we also show that the infinitesimal form of the Weil and Suslin reciprocity laws follow from the residue theorem.

Beginning with the work of Bloch, Gersten, and Quillen (cf. [5] and [16]) one has understood that there is an intricate relationship between K-theory and higher codimensional algebraic cycles. For X an algebraic curve, the Chow group $CH^1(X)$ is defined as the cokernel of the mapping obtained by taking global sections of the surjective mapping of Zariski sheaves

$$\underline{\underline{\mathbb{C}}}(X)^* \xrightarrow{\text{div}} \bigoplus_{x \in X} \underline{\underline{\mathbb{Z}}}_x \to 0. \tag{1.13}$$

This sheaf sequence completes to the exact sequence

$$0 \to \mathcal{O}^*_X \to \underline{\underline{\mathbb{C}}}(X)^* \to \bigoplus_{x \in X} \underline{\underline{\mathbb{Z}}}_x \to 0\,, \tag{1.14}$$

and the exact cohomology sequence gives the well-known identification

(1.15) $$CH^1(X) \cong H^1(\mathcal{O}_X^*).$$

For X an algebraic surface, the analogue of (1.13) is

(1.16) $$\bigoplus_{\left\{\substack{Y \text{ irred} \\ \text{curve}}\right.} \underline{\underline{\mathbb{C}}}(Y)^* \xrightarrow{\text{div}} \bigoplus_{x \in X} \underline{\underline{\mathbb{Z}}}_x \to 0.$$

Whereas the kernel of the map in (1.13) is evidently $\mathcal{O}_X^*$, for (1.16) it is a nontrivial result that the kernel is the image of the map

$$\underline{\underline{K}}_2(\mathbb{C}(X)) \xrightarrow{T} \bigoplus_{\left\{\substack{Y \text{ irred} \\ \text{curve}}\right.} \underline{\underline{\mathbb{C}}}(Y)^*$$

given by the tame symbol. It is at this juncture that K-theory enters the picture in the study of higher codimension algebraic cycles. The sequence (1.16) then completes to the analogue of (1.14), the Bloch-Gersten-Quillen exact sequence

$$0 \to \mathcal{K}_2(\mathcal{O}_X) \to \underline{\underline{K}}_2(\mathbb{C}(X)) \to \bigoplus_{\left\{\substack{Y \text{ irred} \\ \text{curve}}\right.} \underline{\underline{\mathbb{C}}}(Y)^* \to \bigoplus_{x \in X} \underline{\underline{\mathbb{Z}}}_x \to 0$$

which in turn leads to Bloch's analogue

(1.17) $$CH^2(X) \cong H^2(\mathcal{K}_2(\mathcal{O}_X))$$

of (1.15), which opened up a whole new perspective in the study of algebraic cycles.

The infinitesimal form of (1.17) is also due to Bloch (cf. [4] and [27]) with important amplifications by Stienstra [6]. In this work the van der Kallen result is central. Because it is important for our work to understand in detail the infinitesimal properties of the Steinberg relations that give the (Milnor) K-groups, we have in the appendix to chapter 6 given the calculations that lie behind (1.12). At the end of this appendix we have amplified on the above heuristic argument that shows from a geometric perspective how K-theory and absolute differentials necessarily enter into the study of higher codimensional algebraic cycles.

In chapter 7 we give the formal definition

$$\underline{\underline{T}}Z^2(X) = \lim_{\left\{\substack{Z \text{ codim } 2 \\ \text{subscheme}}\right.} \mathcal{E}xt^2_{\mathcal{O}_X}\left(\mathcal{O}_Z, \Omega^1_{X/\mathbb{Q}}\right)$$

for the tangent sheaf to the sheaf of 0-cycles on a smooth algebraic surface X. We show that this is equivalent to the geometric description discussed above. Then, based on a construction of Angéniol and Lejeune-Jalabert [19], we define a map

$$T\mathrm{Hilb}^2(X) \to TZ^2(X),$$

thereby showing that the tangent to an arc in $Z^2(X)$ given as the image in $Z^2(X)$ of an arc in $\mathrm{Hilb}^2(X)$ depends only on the tangent to that arc in $T\mathrm{Hilb}^2(X)$.

In summary, the geometric description has the advantages:

– it is additive;

– it depends only on $z(t)$ as a cycle;

– it depends only on $z(t)$ up to first order in t;

– it has clear geometric meaning.

It is however not clear that two families of effective cycles that represent the same element of $T\,\mathrm{Hilb}^2(X)$ have the same tangent under the geometric description. The formal definition has the properties:

– it clearly factors through $T\mathrm{Hilb}^2(X)$;

– it is easy to compute in examples.

But additivity does not make sense for arbitrary schemes, and in the formal definition it is not clear that $z'(0)$ depends only on the cycle structure of $z(t)$. For this reason it is important to show their equivalence.

In chapter 8 we give the definitions of some related spaces, beginning with the definition

$$\underline{\underline{T}}Z^1(X) = \lim_{\left\{\substack{Z \text{ codim} 1 \\ \text{subscheme}}\right.} \mathcal{E}xt^1_{\mathcal{O}_X}(\mathcal{O}_Z, \mathcal{O}_X)$$

for the sheaf of divisors on a smooth algebraic surface X. Actually, this definition contains interesting geometry not present for divisors on curves. In section 8.2 this geometry is discussed both directly and dually using differential forms and residues. As background for this, we review duality with emphasis on how one may use the theory to compute in examples.

In section 8.2 we give the definition

$$\underline{\underline{T}}Z^1_1(X) = \bigoplus_{\left\{\substack{Y \text{ codim } 1 \\ Y \text{ irred}}\right.} \underline{\underline{H}}^1_y(\Omega^1_{X/\mathbb{Q}})$$

for the tangent sheaf to the Zariski sheaf $\bigoplus_Y \underline{\underline{\mathbb{C}}}(Y)^*$. Underlying this definition is an interesting mix of arithmetic and geometry which is illustrated in a number of examples. With this definition there is a natural map $\underline{\underline{T}}Z^1_1(X) \to \underline{\underline{T}}Z^2(X)$ and passing to global sections we may define the *geometric tangent space* to the Chow group $CH^2(X)$ by

$$T_{\mathrm{geom}}CH^2(X) = TZ^2(X)/\mathrm{image}\left\{TZ^1_1(X) \to TZ^2(X)\right\}.$$

Both the numerator and denominator on the RHS have geometric meaning and are amenable to computation in examples. The main result of this work is then given by the

Theorem: (i) *There is a natural surjective map*

$$\begin{pmatrix} \text{arcs in} \\ \bigoplus_Y \underline{\underline{\mathbb{C}}}(Y)^* \end{pmatrix} \to \underline{\underline{T}}Z^1_1(X).$$

(ii) *Denoting by* $T_{\mathrm{formal}}CH^2(X)$ *the formal tangent space to the Chow group given by Bloch* [4], [27], *there is a natural identification*

$$T_{\mathrm{geom}}CH^2(X) \cong T_{\mathrm{formal}}CH^2(X).$$

Contained in (i) and (ii) in this theorem is a geometric existence result, albeit at the infinitesimal level. The interesting but significant difficulties in "integrating" this result are discussed in section 8.4 and again in chapter 10.

In chapter 9 we give some applications and examples. Classically, on an algebraic curve Abel's differential equations—by which we mean the infinitesimal form of Abel's theorem—express the infinitesimal constraints that a 0-cycle move in a rational equivalence class. An application of our work gives an extension of Abel's differential equations to 0-cycles on an n-dimensional smooth variety X. For X a regular algebraic surface defined over $\mathbb{Q}$ these conditions take the following form: Let $z = \sum_i x_i$ be a 0-cycle, where for simplicity of exposition we assume that the x_i are distinct. Given $\tau_i \in T_{x_i}X$ we ask *when is*

$$\tau = \sum_i (x_i, \tau_i) \in TZ^2(X) \tag{1.18}$$

tangent to a rational equivalence? Here there are several issues that one does not see in the curve case. One is that because of the cancellation phenomenon in higher codimension discussed above it is essential to allow creation/annihilation arcs in $Z^2(X)$, so it is understood that a picture like

$$\begin{array}{c} x_-(t) \\ \leftarrow \cdot \rightarrow \\ x_+(t) \end{array}$$

$$\begin{cases} x(t) = x_+(t) - x_-(t), \\ x_+(0) = x_-(0), \\ x'_+(0) = -x'_-(0) \end{cases}$$

is allowed, and a picture like

$$\overset{\bullet}{x} \longrightarrow \tau$$

could be the tangent to a simple arc $x(t)$ in X with $x(0) = x$ and $x'(0) = \tau$, or it could be the tangent to an arc

$$z(t) = x_1(t) + x_2(t) - x_3(t),$$

where

$$\begin{cases} x_1(0) = x_2(0) = x_3(0) = x, \\ x'_1(0) + x'_2(0) - x'_3(0) = \tau, \end{cases}$$

and so forth.[2]

Second, we can only require that τ be tangent to a first order arc in $Z^2_{\text{rat}}(X)$. Alternatively, we could require (i) that τ be tangent to a formal arc in $Z^2_{\text{rat}}(X)$, or

[2]Of course, for curves one may introduce creation/annihilation arcs, but as noted in point (e) above it is only in higher codimension, due to the presence of "irrelevant" rational equivalences, that they play an essential role.

(ii) that τ be tangent to a geometric arc in $Z^2_{\mathrm{rat}}(X)$. There are heuristic geometric reasons that (i) may be the same as tangent to a first-order arcs, but although (ii) *may* be equivalent for 0-cycles on a surface (essentially Bloch's conjecture), there are Hodge-theoretic reasons why for higher dimensional varieties the analogue of (ii) cannot in general be equivalent to tangency to a first order rational equivalence for general codimension 2 cycles (say, curves on a threefold). In any case, there are natural pairings

$$\langle\ ,\ \rangle : \Omega^2_{X/\mathbb{Q},x} \otimes T_x X \to \Omega^1_{\mathbb{C}/\mathbb{Q}} \tag{1.19}$$

and the condition that (1.18) be tangent to a first order rational equivalence class is

$$\langle \omega, \tau \rangle =: \sum_i \langle \omega, \tau_i \rangle = 0 \qquad \text{in} \quad \Omega^1_{\mathbb{C}/\mathbb{Q}} \tag{1.20}$$

for all $\omega \in H^0(\Omega^2_{X/\mathbb{Q}})$. If the $x_i \in X(k)$ then the pairing (1.19) lies in $\Omega^1_{k/\mathbb{Q}}$.

At one extreme, if $z = \sum_i x_i \in Z^2\big(X(\bar{\mathbb{Q}})\big)$ then all $\langle \omega, \tau_i \rangle = 0$ and the main theorem stated above gives a geometric existence result which is an infinitesimal version of the conjecture of Bloch-Beilinson [22]. At the other extreme, taking the x_i to be independent transcendentals we obtain a quantitative version of the theorem of Mumford-Roitman (cf. [1] and [2]). In between, the behavior of how a 0-cycle moves infinitesimally in a rational equivalence class is very reminiscent of the behavior of divisors on curves where $h^{2,0}(X)$ *together with* tr deg (k) play the role of the genus of the curve.

In section 9.2 we discuss the integration of Abel's differential equations. The exact meaning of "integration" is explained there—roughly it means defining a Hodge-theoretic object $\mathcal{H}$ and map

$$\psi : Z^n(X) \to \mathcal{H} \tag{1.21}$$

whose codifferential factors through the map

$$T^*CH^n(X) \to T^*Z^n(X).$$

For curves, denoting by $Z^1(X)_0$ the divisors of degree zero, the basic classical construction is the pairing

$$H^0(\Omega^1_{X/\mathbb{C}}) \otimes Z^1(X)_0 \to \mathbb{C} \text{ mod periods}$$

given by

$$\omega \otimes z \xrightarrow{\psi} \int_\gamma \omega, \qquad \partial\gamma = z.$$

As z varies along an arc z_t

$$\frac{d}{dt}\big(\psi(\omega \otimes z_t)\big) = \langle \omega, z' \rangle$$

where the right-hand side is the usual pairing

$$H^0(\Omega^1_{X/\mathbb{C}}) \otimes TZ^1(X) \to \mathbb{C}$$

of differential forms on tangent vectors. This of course suggests that the usual abelian sums should serve to integrate Abel's differential equations in the case of curves.

In [32] we discussed the integration of Abel's differential equations in general. Here we consider the first nonclassical case of a regular surface X defined over $\mathbb{Q}$, and we shall explain how the geometric interpretation of (1.20) suggests how one may construct a map (1.21) in this case. What is needed is a pairing

$$H^0(\Omega^2_{X/\mathbb{C}}) \otimes Z^2(X)_0 \to \int_\Gamma \omega \quad \text{mod periods,} \tag{1.22}$$

where Γ is a (real) 2-dimensional chain that is constructed from z using the assumptions that $\deg z = 0$ and that X is regular. If $z \in Z^2\big(X(k)\big)_0$, then using the spread construction together with (1.10) we have a pairing

$$H^0(\Omega^2_{X/\mathbb{Q}}) \otimes T_z Z^2(X)_0 \to T^*_{s_0} S, \tag{1.23}$$

which if we compare it with (1.10) will, according to (1.19) and (1.20), give the conditions that z move infinitesimally in a rational equivalence class. Writing (1.23) as a pairing

$$H^0(\Omega^2_{X/\mathbb{Q}}) \otimes TZ^2(X) \otimes TS \to \mathbb{C}$$

suggests in analogy to the curve case that in (1.21) the 2-chain Γ should be traced out by 1-chains γ_s in X parametrized by a curve λ in S. Choosing γ_s so that $\partial\gamma_s = z_s$ and taking for λ a closed curve in S, we are led to set $\Gamma = \bigcup_{s\in\lambda} \gamma_s$ and define for $\omega \in H^0(\Omega^2_{X/\mathbb{C}})$

$$I(z, \omega, \lambda) = \int_\Gamma \omega \quad \text{mod periods.} \tag{1.24}$$

As is shown in section 9.2 this gives a *differential character* on S that depends only on the k-rational equivalence class of z.[3] If one *assumes* the conjecture of Bloch and Beilinson, then the triviality of $I(z, \cdot, \cdot)$ implies that z is rationally equivalent to zero; this would be an analogue of Abel's theorem for 0-cycles on a surface.

In section 9.3 we give explicit computations for surfaces in $\mathbb{P}^3$ leading to the following results:

Let X be a general surface in $\mathbb{P}^3$ of degree $d \geqq 5$. Then, for any point $p \in X$

$$T_p X \cap TZ^2(X)_{\text{rat}} = 0.$$

If $d \geqq 6$, then for any distinct points $p, q \in X$

$$(T_p X + T_q X) \cap TZ^2(X)_{\text{rat}} = 0.$$

The first statement implies that a general X contains no rational curve—, that is, a g^1_1—which is a well-known result of Clemens. The second statement implies that a general X of degree $\geqq 6$ does not contain a g^1_2. It may well be that the method of proof can be used to show that for each integer k there is a $d(k)$ such that for $d \geqq d(k)$ a general X does not contain a g^1_k.

In section 9.4 we discuss what seems to be the only nonclassical case where the Chow group is explicitly known: the isomorphism

$$Gr^2 CH^2(\mathbb{P}^2, T) \cong K_2(\mathbb{C}) \tag{1.25}$$

---

[3]The regularity of X enters in the rigorous construction and in the uniqueness of the lifting of ω to $H^0(\Omega^2_{X/\mathbb{Q}})$. Also, the construction is only well-defined modulo torsion. Finally, as discussed in section 9.3, one must "enlarge" the construction (1.24) to take into account all the transcendental part $H^2(X)_{\text{tr}}$ of the second cohomology group of X.

due to Bloch and Suslin [26], [21]. We give a proof of (1.25) similar to that of Totaro [9], showing that it is a consequence of the Suslin reciprocity law together with elementary geometric constructions. This example was of particular importance to us as it was one where the infinitesimal picture could be understood explicitly. In particular, we show that if tr deg $k = 1$ so that S is an algebraic curve, the invariant (1.24) coincides with the regulator and the issue of whether it captures rational equivalence, modulo torsion, reduces to an analogue of a well-known conjecture about the injectivity of the regulator.

In the last chapter we discuss briefly some of the larger issues that this study has raised. One is whether or not the space of codimension p cycles $Z^p(X)$ is at least "formally reduced." That is, given a tangent vector $\tau \in TZ^p(X)$, is there a formal arc $z(t)$ in $Z^p(X)$ with tangent τ? If so, is $Z^p(X)$ "actually reduced"; i.e., can we choose $z(t)$ to be a geometric arc? Here we are assuming that a general definition of $TZ^p(X)$ has been given extending that given in this work when $p = 1$ and $p = n$, and that there is a natural map

$$T\mathrm{Hilb}^P(X) \to TZ^p(X).$$

The first part of the following proposition is proved in this book; the second is a result of Ting Fei Ng [39], the idea of whose proof is sketched in chapter 10:

(1.26) $\qquad Z^p(X)$ *is reduced for* $p = n, 1$.

What this means is that for $p = n, 1$ every tangent vector in $TZ^p(X)$ is the tangent to a geometric arc in $Z^p(X)$. For $p = n$ this is essentially a local result. However, for $p = 1$ and $n \geqq 2$ it is well known that $\mathrm{Hilb}^1(X)$ may not be reduced. Already when $n = 2$ there exist examples of a smooth curve Y in an algebraic surface and a normal vector field $\nu \in H^0(N_{Y/X})$ which is not tangent to a geometric definition of Y in X; i.e., ν may be obstructed. However, when we consider Y as a codimension one cycle on X the above result implies that there is an arc $Z(t)$ in $Z^1(X)$ with

$$\begin{cases} Z(0) = Y, \\ Z'(0) = \nu; \end{cases}$$

in particular, *allowing Y to deform as a cycle kills the obstructions.*

For Hodge-theoretic reasons, (1.26) cannot be true in general—as discussed in chapter 10, when $p = 2$ and $n = 3$ the result is not true. Essentially there are two possibilities:

(i) $Z^p(X)$ *is not reduced.*

(ii) $Z^p(X)$ *is formally, but not actually, reduced.*

Here we are using "reduced" as if $Z^p(X)$ had a scheme structure, which of course it does not. What is meant is that first an mth order arc is given by a finite linear combination of the map to the space of cycles induced by maps

$$\mathrm{Spec}\left(\mathbb{C}[t]/t^{m+1}\right) \to \mathrm{Hilb}^P(X), \qquad m \geqq 1.^4$$

[4] The issue of the equivalence relation on such maps to define the same cycle is non-trivial — cf. section 10.2. In fact, the purpose of chapter 10 is to raise issues that we feel merit further study.

The tangent to such an arc factors as in

$$\begin{array}{ccc} \mathrm{Spec}(\mathbb{C}[t]/t^{m+1}) & \to & Z^p(X) \\ \downarrow & \searrow & \\ \mathrm{Spec}(\mathbb{C}[t]/t^2) & \to & TZ^p(X) \end{array}$$

where the top row is the finite linear combination of the above maps, and where the bottom row is surjective. To say that $\tau \in TZ^p(X)$ is *unobstructed to order* m means that it is in the image of the dotted arrow. To say that it is *formally reduced* means that it is unobstructed to order m for all m. To say that it is *actually reduced* means that it comes from a geometric arc

$$B \to Z^p(X).$$

Another anomaly of the space of cycles is the presence of *null curves* in the Chow group, these being curves $z(t)$ in $CH^p(X)$ that are nonconstant but whose derivative is identically zero. They arise from tangent vectors to rational equivalences that do not arise from actual rational equivalences (nonreduced property of $TZ^n_{\mathrm{rat}}(X) =:$ image $\{TZ^n_1(X) \to TZ^n(X)\}$—see below for notation). Thus, if one thinks in the language of differential equations:

(1.27) *Because of the presence of null curves, there can be no uniqueness in the integration of Abel's differential equations.*

Thus, the usual existence and uniqueness theorems of differential equations both fail in our context. Heuristic considerations suggest that one must add additional arithmetic considerations to have even the possibility of convergent iterative constructions. The monograph concludes with a discussion of this issue in section 10.4.

We have used classical terminology in discussing the spaces of cycles on an algebraic variety, as if the $Z^p(X)$ were themselves some sort of variety. However, because of properties such as (1.26) and (1.27) the $Z^p(X)$ are decidedly nonclassical objects. This nonclassical behavior is combined Hodge-theoretic and arithmetic in origin, and in our view understanding it presents a deep challenge in the study of algebraic cycles.

To conclude this introduction we shall give some references and discuss the relationship of this material to some other works on the space of cycles on an algebraic variety.

Our original motivation stems from the work of David Mumford and Spencer Bloch some thirty odd years ago. The paper [Rational equivalence of 0-cycles on surfaces., *J. Math. Kyoto Univ.* **9** (1968), 195–204] by Mumford showed that the story for Chow groups in higher codimensions would be completely different from the classical case of divisors. Certainly one of the questions in our minds was whether Mumford's result and the subsequent important extensions by Roitman [Rational equivalence of zero-dimensional cycles (Russian), *Mat. Zametki* **28**(1) (1980), 85–90, 169] and [The torsion of the group of 0-cycles modulo rational equivalence, *Ann. of Math.* **111** (2) (1980), 553–569] could be understood, and perhaps refined, by defining the tangent space to cycles and then passing to the quotient by infinitesimal rational equivalence—this turned out to be the case.

The monograph *Lectures on algebraic cycles* (Duke University Mathematics Series, IV. Duke University, Mathematics Department, Durham, N.C., 1980. 182 pp.) by Bloch was one of the major milestones in the study of Chow groups and provided significant impetus for this work. The initial paper [K_2 and algebraic cycles, *Ann. of Math.* **99**(2) (1974), 349–379] by Bloch its successor [Algebraic cycles and higher K-theory, *Adv. in Math.* **61**(3) (1986), 267–304] together with Quillen's work [16] brought K-theory into the study of cycles, and trying to understand geometrically what is behind this was one principal motivation for this work. We feel that we have been able to do this infinitesimally by giving a geometric understanding of how absolute differentials necessarily enter into the description of the tangent space to the space of 0-cycles on a smooth variety. One hint that this should be the case came from Bloch's early work [*On the tangent space to Quillen K-theory, Lecture Notes in Math.* **341** (1974), Springer-Verlag] and summarized in [4] and with important extensions by Stienstra, Balere [On K_2 and K_3 of truncated polynomial rings, Algebraic K-theory, Evanston 1980 (Proc. Conf., Northwestern Univ., Evanston, Ill., 1980), pp. 409–455, Lecture Notes in Math. **854**, Springer, Berlin, 1981].

Another principal motivation for us has been provided by the conjectures of Bloch and Beilinson. These are explained in sections 6 and 8 of [Ramakrishnan, Dinakar, Regulators, algebraic cycles, and values of L-functions, *Contemp. Math.* **83** (1989), 183–310] and in [Jannsen, U., Motivic sheaves and filtrations on Chow groups. Motives (Seattle, WA, 1991), 245–302, *Proc. Sympos. Pure Math.* **55**, Part 1, Amer. Math. Soc., Providence, RI, 1994]. Our work provides a geometric understanding and verification of these conjectures at the infinitesimal level, and it also points out some of the major obstacles to "integrating" these results [8].

In an important work, Blaine Lawson introduced a topology on the space $Z^p(X)$ of codimension p algebraic cycles on a smooth complex projective variety. Briefly, two codimension-p cycles z, z' written as

$$\begin{cases} z = z_+ - z_- \\ z' = z'_+ - z'_- \end{cases}$$

where $z_\pm$, $z'_\pm$ and effective cycles are close, if z_+, z'_+ and z_-, z'_- are close in the usual sense of closed subsets of projective space. Lawson then shows that $Z^p(X)$ has the homotopy type of a CW complex, and from this he proceeds to define the Lawson homology of X in terms of the homotopy groups of $Z^p(X)$. His initial work triggered an extensive development, many aspects of which are reported on in his talk at ICM Zürich (cf. [Lawson, Spaces of Algebraic Cycles—Levels of Holomorphic Approximation, *Proc. ICM Zürich*, 574–584] and the references cited therein).

In this monograph, although we do not define a topology on $Z^p(X)$, we do define and work with the concept of a (regular) arc in $Z^p(X)$. Implicit in this is the condition that two cycles z, z' as above should be close: First, there should be a common field of definition for X, z, and z'. This leads to the spreads

$$\begin{array}{c} \mathcal{Z}, \mathcal{Z}' \subset \mathcal{X} \\ \downarrow \\ S \end{array}$$

as discussed in chapter 4 below, and z, z' should be considered close if $\mathcal{Z}$, $\mathcal{Z}' \in Z^p(\mathcal{X})$ are close in the Lawson sense (taking care to say what this means, since the spread is not uniquely defined). As seen in the diagram,

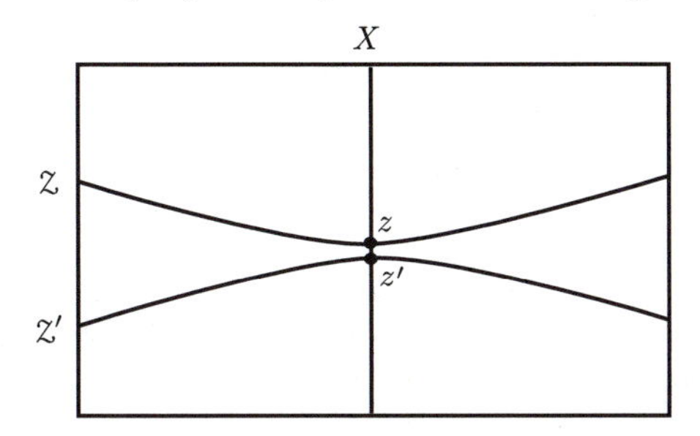

two cycles may be Lawson close without being close in our sense. We do not attempt to formalize this, but rather wish only to point out one relationship between the theory here and that of Lawson and his coworkers.

Finally, we mention that some of the early material in this study has appeared in [23].

Mark Green wishes to acknowledge the National Science Foundation for supporting this research over a period of years. Both authors are grateful to Sarah Warren for a wonderful job of typing a difficult manuscript.

Chapter Two

The Classical Case When $n = 1$

We begin with the case $n = 1$, which is both suggestive and in some ways misleading. Most of the material in this chapter is standard but will help to motivate what comes later. We want to define the tangent to an arc

$$z(t) = \sum_i n_i x_i(t)$$

in the space $Z^1(X)$ of 0-cycles on a smooth algebraic curve X. Later on we will more precisely define what we mean by such an arc—for the moment one may think of the $x_i(t)$ as being given in local coordinates by a Puiseaux series in t.

The *degree*

$$\deg z(t) =: \sum_i n_i$$

is constant in t. Among all arcs $z(t)$, those with

$$\begin{cases} z(0) = 0 \\ z(t) \not\equiv 0 \end{cases}$$

are of particular interest. The presence of such arcs is one main difference between the configuration spaces $X^{(m)}$—sets of m points on X—and the space $Z^1(X)$ of 0-cycles. We may write such an arc as sum of arcs

$$z(t) = z^+(t) - z^-(t)$$

where $z^{\pm}(t)$ are arcs in $X^{(m)}$ with $z^+(0) = z^-(0)$, and we think of $z(t)$ as a *creation/annihilation* arc.

We denote by $|z(t)|$ the support of the 0-cycle $z(t)$, and for $x \in X$ denote by $Z^1_{\{x\}}(X)$ the set of arcs $z(t)$ with

$$\lim_{t \to 0} |z(t)| = x.$$

It will suffice to define the tangent space $T_{\{x\}} Z^1(X)$ to arcs in $Z^1_{\{x\}}(X)$. From a sheaf-theoretic perspective, in the Zariski topology we denote by $\underline{\underline{Z}}^1(X)$ the Zariski sheaf of 0-cycles on X, and by $\underline{\underline{T}} Z^1(X)$ the to be defined tangent sheaf to $Z^1(X)$. The stalks of $\underline{\underline{T}} Z^1(X)$ are then given by

$$\underline{\underline{T}} Z^1(X)_x = T_{\{x\}} Z^1(X).$$

In fact, the sheaf-theoretic perspective suggests one possible definition of $T_{\{x\}}Z^1(X)$. Namely, we have the standard exact sheaf sequence

$$0 \to \mathcal{O}_X^* \to \underline{\underline{\mathbb{C}}}(X)^* \xrightarrow{\text{div}} \underline{\underline{Z}}^1(X) \to 0. \tag{2.1}$$

With any reasonable definition the tangent sheaf to $\mathcal{O}_X^*$ is $\mathcal{O}_X$, with the map being

$$\text{tangent to } \{f + tg + \cdots\} = g/f$$

where $f \in \mathcal{O}_{X,x}^*$ and $g \in \mathcal{O}_{X,x}$, and similarly the tangent sheaf to $\underline{\underline{\mathbb{C}}}(X)^*$ should be $\underline{\underline{\mathbb{C}}}(X)$. This suggests that the tangent sheaf sequence to (2.1) can be defined and should be the well-known sequence

$$0 \to \mathcal{O}_X \to \underline{\underline{\mathbb{C}}}(X) \to \mathcal{PP}_X \to 0 \tag{2.2}$$

where $\mathcal{PP}_X$ is the sheaf of principal parts. Thus, we should at least provisionally define

$$\underline{\underline{T}}Z^1(X) = \mathcal{PP}_X.$$

More explicitly, we consider an arc

$$f(t) = f + tg + \cdots$$

in $\underline{\underline{\mathbb{C}}}(X)_x^*$ where $f \in \mathbb{C}(X)^*$ and $g \in \mathbb{C}(X)$. Then

$$z(t) = \text{div } f(t)$$

is an arc in $Z^1(X)$, which we assume to be in $Z^1_{\{x\}}(X)$, and with the above provisional definition the tangent to $z(t)$ is given by

$$z'(0) = [g/f]_x$$

where $[g/f]_x$ is the principal part at x of the rational function g/f. More formally

Definition: Two arcs $z(t)$ and $\widetilde{z}(t)$ in $Z^1_{\{x\}}(X)$ are said to be *equivalent to first order*, written

$$z(t) \equiv_{1^{\text{st}}} \widetilde{z}(t), \tag{2.3}$$

if with the obvious notation we have

$$[g/f]_x = [\widetilde{g}/\widetilde{f}]_x.$$

The tangent space to $Z^1_{\{x\}}(X)$ is then provisionally defined by

$$T_{\{x\}}Z^1(X) = Z^1_{\{x\}}(X)/\equiv_{1^{\text{st}}}. \tag{2.4}$$

This is not the formal definition, which will be given later and will be shown to be equivalent to the provisional definition.

Note: Below we will explicitly write out what this all means. There it will be seen that (2.3) is equivalent to having

$$\widetilde{z}(t) = \text{div}\,(f + tg + \cdots),$$

i.e., we may take $\widetilde{f} = f$ and $\widetilde{g} = g$.

Perhaps the most natural notion of first order equivalence to use is the tangent space to the Hilbert scheme. A family of effective 0-cycles $z(t)$ with $z(0) = z$ gives a map

$$\begin{aligned} \mathcal{I}_z &\to \mathcal{O}_X/\mathcal{I}_z, \\ f &\mapsto \frac{df}{dt}. \end{aligned}$$

For families of effective 0-cycles $z_1(t), z_2(t)$ with

$$z_1(0) = z_2(0) = z$$

one says

$$z_1(t) \widetilde{\equiv}_{1^{\text{st}}} z_2(t)$$

if they induce the same map $\mathcal{I}_z \to \mathcal{O}_X/\mathcal{I}_z$. The subgroup of effective arcs in $Z^1(X)$ starting at a z with $|z| = x$ generates an equivalence relation $\widetilde{\equiv}_{1^{\text{st}}}$, which will be seen to be the same as that given by Definition (2.4). Thus, for 0-cycles on a curve, if we denote by $\mathrm{Hilb}^1_k(X)$ the 0-dimensional subschemes of degree k we have that

$$T\mathrm{Hilb}^1_k(X) \to TZ^1(X)$$

injects and

$$\lim_{k\to\infty} T_{kx}\mathrm{Hilb}^1_k(X) \cong T_{\{x\}}Z^1(X).$$

This will *not* be the case in higher codimension.

Classically, the tangent to an arc in the space of divisors on an algebraic curve appeared in the theory of abelian sums (see below). This suggests that the dual space to $T_{\{x\}}Z^1(X)$ should be related to differential forms. In fact there is a nondegenerate pairing

$$\mathcal{PP}_{X,x} \otimes_{\mathbb{C}} \Omega^1_{X/\mathbb{C},x} \to \mathbb{C} \tag{2.5}$$

given by

$$\tau \otimes \omega \to \mathrm{Res}_x(\tau\omega).$$

Here, in terms of a local uniformizing parameter ξ centered at x, we are thinking of $\mathcal{PP}_{X,x}$ as the space of finite Laurent tails

$$\mathcal{PP}_{X,x} \cong \left\{ \tau = \sum_{k=1}^{N} \frac{a_k}{\xi^k} \right\}.$$

For $\omega = (\sum_{\ell \geqq 0} b_\ell \xi^\ell) d\xi$

$$\mathrm{Res}_x(\tau\omega) = \sum_{\ell \geqq 0} a_{\ell+1} b_\ell,$$

from which we see that the pairing (2.5) is nondegenerate. A consequence is the natural identification of the provisional tangent space

$$T_{\{x\}}Z^1(X) \cong \mathrm{Hom}^c_{\mathbb{C}}(\Omega^1_{X/\mathbb{C},x}, \mathbb{C}), \tag{2.6}$$

where $\mathrm{Hom}^c_{\mathbb{C}}(\cdot, \cdot)$ denotes the continuous $\mathbb{C}$-linear homomorphisms; i.e., those φ that annihilate $\mathfrak{m}_x^N \Omega^1_{X/\mathbb{C},x}$ for some $N = N(\varphi)$.

Turning to abelian sums, we consider an arc of effective 0-cycles

$$z(t) = \sum_{i=1}^{m} x_i(t), \qquad x_i(0) = x$$

in $Z^1_{\{x\}}(X)$, given by a regular mapping

$$B \to X^{(m)}$$

from an algebraic curve B with local uniformizing parameter t into the m-fold symmetric product of X. Here, there is a reference point $b_0 \in B$ with $t(b_0) = 0$. The corresponding abelian sum is

$$\sum_i \int_{x_0}^{x_i(t)} \omega.$$

It is well known that this abelian sum is regular in t for t near 0, and we set

$$I(z, \omega) = \frac{d}{dt}\left(\sum_i \int_{x_0}^{x_i(t)} \omega \right)_{t=0}. \tag{2.7}$$

Proposition: *Denoting by $z'(0)$ the Laurent tail corresponding to the tangent vector to $z(t)$, we have*

$$I(z, \omega) = \operatorname{Res}_x\big(z'(0)\omega\big). \tag{2.8}$$

Thus, the first order terms in abelian sums give an alternate description of the isomorphism (2.6).

Proof: If

$$z(t) = \text{ divisor of } f + tg$$

then we need to show that

$$\operatorname{Res}_x\big((g/f)\omega\big) = I(z, \omega).$$

This follows from the general

Lemma: *Expressed in terms of a complex variable x, if we write $f + tg = \prod_{i=1}^m (x - x_i(t)) \mod o(t^{1+\epsilon})$ and*

$$\omega = x^k dx$$

then

$$\operatorname{Res}_0\left(\frac{g\omega}{f}\right) = -\lim_{t \to 0} \sum_{i=1}^{m} x_i(t)^k x_i'(t).$$

Proof: We give two independent proofs.

First proof:

$$f + tg = x^m + \sum_{j=1}^{m} (-1)^j \sigma_j\big(x_1(t), \ldots, x_m(t)\big) x^{m-j} \qquad \bmod\ O(t^{1+\epsilon}).$$

So

$$g(x) = \sum_{j=1}^{m} (-1)^j \frac{d}{dt} \sigma_j\big(x_1(t), \ldots, x_m(t)\big) x^{m-j}\Big|_{t=0}$$

and thus

$$\mathrm{Res}_0\left(\frac{g\omega}{f}\right) = (-1)^{k+1} \frac{d}{dt} \sigma_{k+1}\big(x_1(t), \ldots, x_m(t)\big)\Big|_{t=0}.$$

By Newton's identities,

$$\sum_{i=1}^{m} x_i(t)^{k+1} = -(k+1)\sigma_{k+1}\big(x_1(t), \ldots, x_m(t)\big) + \text{terms of deg} \geq 2 \text{ in } \sigma_1, \ldots, \sigma_m.$$

Thus

$$\sum_{i=1}^{m} x_i(t)^k x_i'(t) = -\frac{d}{dt}\sigma_{k+1}\big(x_1(t), \ldots, x_m(t)\big) + \sum_{i=1}^{m} p_i(\sigma_1, \ldots, \sigma_m)\sigma_i\big(x_1(t), \ldots, x_m(t)\big),$$

where every term of p_i has positive total degree in σ_i.

Since $\sigma_j\big(x_1(0), \cdots x_i(0)\big) = 0$ for all $j = 1, \ldots, m$, it follows that $p_i(\sigma_1, \cdots, \sigma_i) = 0$ at $t = 0$. So

$$\lim_{t \to 0} \sum_i x_i(t)^k x_i'(t) = -\frac{d}{dt}\sigma_{k+1}(x_1(t), \ldots, x_m(t))\big|_{t=0} = -\mathrm{Res}_0\left(\frac{g\omega}{f}\right).$$

Second proof:

$$\begin{aligned}
\mathrm{Res}_0\left(\frac{g\omega}{f}\right) &= \lim_{t \to 0} \sum_{i=1}^{m} \mathrm{Res}_{x_i(t)}\left(\frac{g\omega}{f + tg}\right) \\
&= \lim_{t \to 0} \sum_{i=1}^{m} \mathrm{Res}_{x_i(t)} \frac{g x^k dx}{\prod_{j-1}^{m}(x - x_j(t))} \\
&= \lim_{t \to 0} \sum_{i=1}^{m} \frac{g(x_i(t)) x_i(t)^k}{\prod_{j \neq i}(x_i(t) - x_j(t))}.
\end{aligned}$$

Differentiating

$$f + tg = \prod_{i=1}^{m}(x - x_i(t))$$

with respect to t gives

$$g(x) = \sum_{i=1}^{m} -x_i'(t) \cdot \prod_{j \neq i} \big(x - x_j(t)\big)$$

so

$$g\big(x_i(t)\big) = -x_i'(t) \cdot \prod_{j \neq i} \big(x_i(t) - x_j(t)\big),$$

and plugging into the earlier formula gives

$$\mathrm{Res}_0\left(\frac{g\omega}{f}\right) = -\lim_{t \to 0} \sum_i x_i(t)^k x_i'(t). \qquad \square$$

For later use it is instructive to examine the abelian sum approach in local analytic coordinates, which we may take to be an analytic disc $\{b : |t(b)| < \delta\}$, where t is a local uniformizing parameter on X and $\delta > 0$ is a positive constant. Call an arc in $X^{(m)}$

$$z(t) = \sum_{i=1}^{m} x_i(t), \quad x_i(0) = x$$

irreducible if $x_i(t) \neq x_j(t)$ for $i \neq j$ and $t \neq 0$, and if analytic continuation around $t = 0$ permutes the $x_i(t)$ transitively. Any arc in $Z^1_{\{x\}}(X)$ is a $\mathbb{Z}$-linear combination of constant arcs and irreducible arcs. Since the tangent map is additive on arcs, it suffices to consider the tangents to irreducible arcs.

An irreducible arc is represented by a convergent Puiseaux series

$$x_i(t) = \sum_{j=0}^{m} a_j \epsilon^{ij} t^{j/m} + O(t^{1+1/m}),$$

where $\epsilon = e^{2\pi\sqrt{-1}/m}$. Then

$$\begin{aligned}\frac{d}{dt}\left(\sum_i \int_x^{x_i(t)} \xi^k d\xi\right)\bigg|_{t=0} &= \sum_{i=1}^{m}\left(\sum_{j=1}^{m} a_j \epsilon^j t^{j/m}\right)^k \sum_j j a_j \epsilon^j t^{(j/m)-1}\bigg|_{t=0} \\ &= \sum_{\left\{\substack{j_1+\cdots+j_{k+1}=m \\ 0 \leqq j_1,\cdots,j_{k+1} \leqq m \\ j}\right.} j_{k+1} a_j a_{j_1} \cdots a_{j_{k+1}}.\end{aligned}$$

This calculation tells us several things. One is that it establishes directly the nondegeneracy of the pairing

$$T_{\{x\}} Z^1(X) \otimes_{\mathbb{C}} \Omega^1_{X/\mathbb{C},x} \to \mathbb{C}$$

given by

$$z \otimes \omega \to I(z, \omega)$$

in (2.7) above. The second is that the multiplicity m of an irreducible arc is uniquely determined by

$$I(z, \mathfrak{m}_x^k \Omega^1_{X/\mathbb{C},x}) = 0, \qquad k \geqq m.$$

Finally, it is instructive to illustrate the proof of the above proposition and the Puiseaux series calculation in the simplest nontrivial case.

Example: A Puiseaux expansion for $m = 2$ is given by

$$\begin{aligned} x_1(t) &= a_1 t^{1/2} + a_2 t + \cdots \\ x_2(t) &= -a_1 t^{1/2} + a_2 t + \cdots \end{aligned}$$

and is defined by $\operatorname{div}(f + tg)$, where

$$f + tg = \xi^2 + t(-2a_2\xi - a_1^2) + o(t^{1+\epsilon})$$

for some $\epsilon > 0$. If we set

$$z(t) = \operatorname{div}(f + tg)$$

then

$$I(z, d\xi) = \lim_{t \to 0} \sum_{i=1}^{2} \xi_i'(t) = 2a_2$$

$$I(z, \xi d\xi) = \lim_{t \to 0} \sum_{i=1}^{2} \xi_i(t)\xi_i'(t) = a_1^2.$$

Note that

$$\operatorname{Res}_0\left(\frac{g d\xi}{f}\right) = -2a_2$$

$$\operatorname{Res}_0\left(\frac{g\xi d\xi}{f}\right) = -a_1^2.$$

Anticipating future discussions, we observe that there is a natural identification

$$\mathcal{PP}_{X,x} \simeq \lim_{i \to \infty} \mathcal{E}xt^1_{\mathcal{O}_{X,x}}(\mathcal{O}_{X,x}/\mathfrak{m}_x^i, \mathcal{O}_{X,x}).$$

To describe this, we resolve

$$\begin{array}{ccccccc} 0 \to \mathcal{O}_{X,x} & \xrightarrow{g^i} & \mathcal{O}_{X,x} & \to & \mathcal{O}_{X,x}/\mathfrak{m}_x^i & \to & 0 \\ \| & & \| & & & & \\ E_1 & & E_0 & & & & \end{array}$$

where g is a local defining equation for x and

$$\operatorname{Hom}_{\mathcal{O}_{X,x}}(E_1, \mathcal{O}_{X,x}) \simeq \mathcal{O}_{X,x}$$

while

$$(g^i)^*\mathrm{Hom}_{\mathcal{O}_{X,x}}(E_0, \mathcal{O}_{X,x}) \subseteq \mathrm{Hom}_{\mathcal{O}_{X,x}}(E_1, \mathcal{O}_{X,x})$$

is equal to

$$g^i \mathcal{O}_{X,x}.$$

So if

$$f \in \mathrm{Hom}_{\mathcal{O}_{X,x}}(E_1, \mathcal{O}_{X,x})$$

then we may identify

$$[f] \in \mathcal{E}xt^1_{\mathcal{O}_{X,x}}(\mathcal{O}_{X,x}/\mathfrak{m}_x^i, \mathcal{O}_{X,x}) \leftrightarrow \frac{f}{g^i} \in \mathcal{PP}_{X,x}.$$

The natural map for $j > i$,

$$\mathcal{E}xt^1_{\mathcal{O}_{X,x}}(\mathcal{O}_{X,x}/\mathfrak{m}_x^i, \mathcal{O}_{X,x}) \to \mathcal{E}xt^1_{\mathcal{O}_{X,x}}(\mathcal{O}_{X,x}/\mathfrak{m}_x^j, \mathcal{O}_{X,x})$$

takes

$$f \longmapsto fg^{j-i}$$

and since

$$\frac{f}{g^i} = \frac{fg^{j-i}}{g^j}$$

we get

$$\lim_{i\to\infty} \mathcal{E}xt^1_{\mathcal{O}_{X,x}}(\mathcal{O}_{X,x}/\mathfrak{m}_x^i, \mathcal{O}_{X,x}) \simeq \mathcal{PP}_{X,x}.$$

We remark that this limit may be expressed using *local cohomology* as

$$\lim_{i\to\infty} \mathcal{E}xt^1_{\mathcal{O}_{X,x}}(\mathcal{O}_{X,x}/\mathfrak{m}_x^i, \mathcal{O}_{X,x}) \cong H^1_{\mathfrak{m}_x}(\mathcal{O}_{X,x}) \cong H^1_x(\mathcal{O}_X).$$

With this in mind we give the formal

Definition: We define the tangent sheaf

$$\underline{\underline{T}}Z^1(X)$$

to be

$$\underline{\underline{T}}Z^1(X) = \bigoplus_{x\in X} \lim_{k\to\infty} \mathcal{E}xt^1_{\mathcal{O}_X}\left(\mathcal{O}_X/\mathfrak{m}_x^k, \mathcal{O}_X\right).$$

We observe that the tangent map

$$\text{arcs in } Z^1_{\{x\}}(X) \to \underline{\underline{T}}Z^1(X)_x$$

is surjective.

The tangent sequence to (2.1) can then be defined and is the exact sheaf sequence

$$0 \to \mathcal{O}_X \to \underline{\underline{\mathbb{C}}}(X) \to \underline{\underline{T}}Z^1(X) \to 0. \tag{2.9}$$

Setting

$$TZ^1(X) = H^0\big(\underline{\underline{T}}Z^1(X)\big)$$

the exact cohomology sequence of (2.9) is

$$\mathbb{C}(X) \xrightarrow{\rho} TZ^1(X) \to H^1(\mathcal{O}_X) \to 0. \tag{2.10}$$

This is the tangent sequence to the exact sequence

$$\mathbb{C}(X)^* \xrightarrow{\text{div}} Z^1(X) \to \text{Pic}(X) \to 1. \tag{2.11}$$

It follows from (2.10) and (2.11), first, that

$$\text{image } \rho = TZ^1_{\text{rat}}(X)$$

is the tangent space to the subgroup $Z^1_{\text{rat}}(X) \subset Z^1(X)$ of 0-cycles that are rationally equivalent to zero. Second, we have

$$\begin{aligned} TZ^1(X)/TZ^1_{\text{rat}}(X) &\cong H^1(\mathcal{O}_X) \\ &\cong TCH^1(X). \end{aligned} \tag{2.12}$$

Now we have a natural map

$$H^1_x(\mathcal{O}_X) \to H^1(\mathcal{O}_X)$$

from local to global cohomology. The diagram

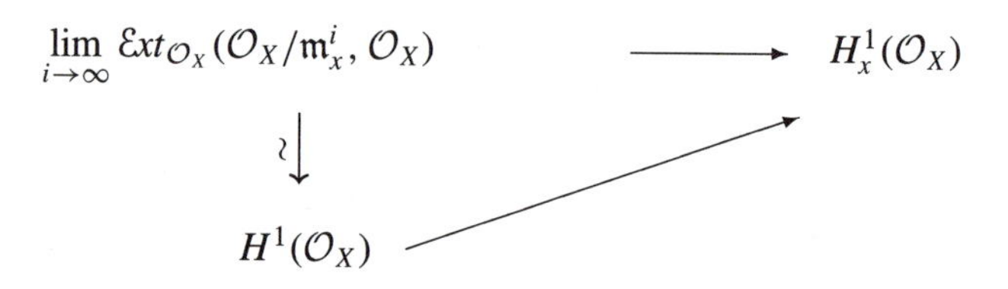

brings together these two maps.

It is this picture that we want to generalize. The description

$$T_{\{x\}}Z^1(X) \cong \text{Hom}^c_{\mathbb{C}}(\Omega^1_{X/\mathbb{C},x}, \mathbb{C})$$

generalizes to 0-cycles on an n-dimensional smooth variety X, but gives the wrong answer—in particular, the analogue of (2.12) fails to hold.[1] The geometric reasons for this are rather subtle and will be discussed below. The *Ext* definition holds generally, but also in a subtle way. Suffice it to say here that two new but closely related phenomena must enter when $n \geqq 2$:

(i) all the forms $\Omega^q_{X/\mathbb{C},x}$, $1 \leqq q \leqq n$ must be used;

(ii) absolute differentials—as it turns out, $\Omega^n_{X/\mathbb{Q}}/\Omega^n_{\mathbb{C}/\mathbb{Q}}$ — must be used.

Of course, (i) is automatic when $n = 1$. As to (ii), when $n = 1$ the quotient $\Omega^1_{X/\mathbb{Q}}/\Omega^1_{\mathbb{C}/Q}$ is isomorphic to $\Omega^1_{X/\mathbb{C}}$, and thus absolute differentials do not enter the picture in this case.

[1]The generalization is, however, interesting and will be discussed in chapter 5.

Chapter Three

Differential Geometry of Symmetric Products

Let X be a smooth variety of dimension n. We are interested in the geometry of configurations of m points on X, which we represent as effective 0-cycles

$$z = x_1 + \cdots + x_m$$

of degree m. Set theoretically such configurations are given by the m-fold symmetric product

$$X^{(m)} = \underbrace{X \times \cdots \times X}_{m} / \Sigma_m$$

where Σ_m is the group of permutations.

We will be especially concerned with arcs of 0-cycles, given by a regular mapping

$$B \xrightarrow{z} X^{(m)}$$

from a smooth (not necessarily complete) curve into $X^{(m)}$. If t is a local uniformizing parameter on B, such an arc may be thought of as

$$z(t) = x_1(t) + \cdots + x_m(t), \tag{3.1}$$

where, in terms of local uniformizing parameters on X, the $x_i(t)$ are given by Puiseaux series in t. We denote by $Z^n_{\{x\}}(X)$ the space given by $\mathbb{Z}$-linear combinations of arcs $z(t)$ as above where all $x_i(0) = x$; i.e., which satisfy

$$z(0) = mx$$

for some m. For example, if $n = 2$ and ξ, η are local uniformizing parameters on X centered at x, then we will have Puiseaux series expansions convergent for $|t| < \delta$ for some constant $\delta > 0$,

$$\begin{cases} \xi_k = a_1\epsilon^k t^{1/m} + a_2\epsilon^{2k}t^{2/m} + \cdots + a_m t + \cdots \\ \eta_k = b_1\epsilon^k t^{1/m} + b_2\epsilon^{2k}t^{2/m} + \cdots + b_m t + \cdots \end{cases} \tag{3.2}$$

where $\epsilon = e^{2\pi\sqrt{-1}/m}$.

It is well known that for $n \geqq 2$ the symmetric products are singular along the diagonals, and in particular along the principal diagonal $\{mx : x \in X\}$.[1] However, essentially because we are interested in 0-cycles and not 0-dimensional subschemes, at this point we will not need to get into the finer aspects of the various candidates for smooth models of the $X^{(m)}$'s (however, see the discussion in chapter 7 below). What we do need is the concept of the regular differential forms of degree q

$$\Omega^q_{X^{(m)}/\mathbb{C},z}$$

[1] As we are interested in the infinitesimal structure of 0-cycles, the geometry of symmetric products along the diagonals is fundamental. Here, understanding the principal diagonals is sufficient.

at a point $z \in X^{(m)}$. This is given by a regular q-form φ defined on the smooth points in a neighborhood of z and which, for any map $f : Y \to X^{(m)}$ where Y is smooth and $f(Y)$ contains a neighborhood of z, $f^*\varphi$ is a regular q-form on Y (cf. [33]). It can then be shown that

$$\Omega^q_{X^{(m)}/\mathbb{C}.mx} \cong \left(\Omega^q_{X^m/\mathbb{C},(\underbrace{x,\ldots,x}_{m})}\right)^{\Sigma_m} \tag{3.3}$$

For us two basic facts concerning the regular forms on symmetric products are

(3.4) *Every* $\omega \in \Omega^q_{X/\mathbb{C},x}$ *induces in* $\Omega^q_{X^{(m)}/\mathbb{C},mx}$ *a regular form* Tr ω, *called the trace of* ω.

Explicitly, Tr ω is induced from the diagonal form $(\omega, \ldots, \omega)$ on X^m, which being invariant under Σ_m descends to a regular form on the smooth points of $X^{(m)}$. We write symbolically

$$(\mathrm{Tr}\,\omega)(z) = \omega(x_1) + \cdots + \omega(x_m).$$

Among forms on symmetric products traces have a number of special properties, including the one of *heredity*: Fixing $x_0 \in X$ there are natural inclusions

$$X^{(m)} \hookrightarrow X^{(m+1)}$$

and the trace of φ on $X^{(m+1)}$ restricts to Tr φ on $X^{(m)}$. Moreover, under the natural map

$$X^{(m_1)} \times X^{(m_2)} \to X^{(m_1+m_2)}$$

traces pull back to a sum of traces. Thus, traces are especially suitable to be thought of as differential forms on the space of 0-cycles.

A second basic fact is the following result, which has been proved by Ting Fai Ng [39]:

(3.5) $\Omega^*_{X^{(m)}/\mathbb{C},mx}$ *is generated over* $\mathcal{O}_{X^{(m)},mx}$ *by sums of elements of form*

$$\mathrm{Tr}\,\omega_1 \wedge \cdots \wedge \mathrm{Tr}\,\omega_k, \quad \omega_i \in \Omega^{q_i}_{X/\mathbb{C},x}.$$

This is clear when $n = 1$, since in this case $X^{(m)}$ is smooth (see below). However, when $n \geqq 2$ and $m \geqq 2$, due to the singularities of symmetric products along the diagonals one may show that we *must* add generators Tr ω for $\omega \in \Omega^q_{X/\mathbb{C},x}$ and all q with $1 \leqq q \leqq n$ to reach all of $\Omega^*_{X^{(m)}/\mathbb{C},mx}$.

For example, when $n = m = 2$ and ξ, η are local uniformizing parameters and we set $\xi_i = \xi(x_i), \eta_i = \eta(x_i)$,

$$\mathrm{Tr}\; d\xi \wedge d\eta = d\xi_1 \wedge d\eta_1 + d\xi_2 \wedge d\eta_2 \tag{3.6}$$

is *not* generated over $\mathcal{O}_{X^{(2)},2x}$ by traces of 1-forms in $\Omega^1_{X/\mathbb{C},x}$. The traces of 1-forms together with (3.6) do generate $\Omega^2_{X^{(2)}/\mathbb{C},2x}$; for example,

$$d\xi_1 \wedge d\eta_2 + d\xi_2 \wedge d\eta_1 = (\mathrm{Tr}\,d\xi) \wedge (\mathrm{Tr}\,d\eta) - \mathrm{Tr}\,(d\xi \wedge d\eta).$$

We will not use (3.5) in the logical development of the theory, and therefore we shall not reproduce Ng's formal proof. However, we feel that it is instructive to see how the first few special cases go.

Let X be an algebraic curve with local uniformizing parameter ξ centered at a point $x \in X$. On the m-fold cartesian product X^m, let ξ_i denote ξ in the ith coordinate. It follows from the theorem on elementary symmetric functions that every function on X^m invariant under the symmetric group Σ_m is uniquely expressible in terms of

$$\begin{cases} \operatorname{Tr} \xi = \xi_1 + \cdots + \xi_m, \\ \operatorname{Tr} \xi^2 = \xi_1^2 + \cdots + \xi_m^2, \\ \quad \vdots \\ \operatorname{Tr} \xi^m = \xi_1^m + \cdots + \xi_m^m. \end{cases}$$

It is a general fact that

$$\operatorname{Tr}(d\varphi) = d(\operatorname{Tr} \varphi),$$

and this implies (3.5) in the case $n = 1$.

Turning to the case $n = 2$, let X be an algebraic surface and ξ, η local coordinates centered at $x \in X$. If we expand forms on X^m around $(x, \ldots, x) \in X^m$, then Σ_m acts homogeneously and the issue is one of the occurrence of the trivial representation in the homogeneous pieces of total degree $p + q$ in

$$\mathbb{C}[\xi_1, \eta_1, \ldots, \xi_m, \eta_m] \otimes \wedge^q \{d\xi_1, d\eta_1, \ldots, d\xi_m, d\eta_m\}.$$

For example, when $q = 1$ the first nontrivial case is when $p = 1$. The terms in $\xi_i d\xi_j$ and $\eta_i d\eta_j$ follow from the $n = 1$ case. By symmetry it will suffice to consider

$$\varphi = \sum_{ij} a_{ij} \xi_i d\eta_j$$

Clearly $a_{21} = a_{31} = \cdots = a_{m1}$. It follows that

$$\varphi - a_{21} \operatorname{Tr}(\xi) \operatorname{Tr}(d\eta) - (a_{11} - a_{21}) \operatorname{Tr}(\xi d\eta)$$

does not involve $d\eta_1$, and hence again it must be zero.

The next case is when $q = 2$ and $p = 0$. Again, the terms in $d\xi_i \wedge d\xi_j$ and $d\eta_i \wedge d\eta_j$ follow from the $n = 1$ case. Thus we may assume that

$$\varphi = \sum_{i,j} a_{ij} d\xi_i \wedge d\eta_j$$

Clearly $a_{11} = \cdots = a_{mm}$ and $a_{21} = \cdots = a_{m1}$. It follows that

$$\varphi - a_{21} (\operatorname{Tr} d\xi)(\operatorname{Tr} d\eta) - (a_{11} - a_{21}) \operatorname{Tr}(d\xi \wedge d\eta)$$

does not involve $d\eta_1$, and hence it must be zero.

The next case is when $q = 2$ and $p = 1$. As before we are reduced to considering

$$\varphi = \sum_{i,j,k} a_{ijk} \xi_i d\xi_j \wedge d\eta_k.$$

We want to show that φ is expressible in terms of the three 2-forms $(\operatorname{Tr} \xi)(\operatorname{Tr} d\xi) \wedge (\operatorname{Tr} d\eta)$, $(\operatorname{Tr} \xi d\xi) \wedge (\operatorname{Tr} d\eta)$ and $(\operatorname{Tr} d\xi) \wedge (\operatorname{Tr} \xi d\eta)$. Since clearly

$$a_{ij1} = a_{ji1} \qquad \text{for} \qquad 2 \le i, j \le n$$

it follows that only three distinct constants $a_{111}, a_{211}, a_{121}$ appear in terms containing η_1. We may therefore subtract from φ a linear combination of the three trace terms to obtain an invariant expression not containing $d\eta_1$, which must then be zero.

The fact that when $n \geqq 2$, we *must* add traces of higher degree forms to generate $\Omega^*_{X^{(m)}/\mathbb{C},mx}$ is a rather subtle differential geometric fact which reflects essential differences in the correct definition of $T_{\{x\}}Z^n(X)$ between the situation discussed above when $n = 1$ and when $n \geqq 2$.

We will now explain how the pairing (2.7) above extends to give in general

$$Z^n_{\{x\}}(X) \otimes_{\mathbb{C}} \Omega^1_{X/\mathbb{C},x} \to \mathbb{C}, \tag{3.7}$$

which is nondegenerate in the second factor. The basic fact is that an arc (3.1) pulls back a regular 1-form in $\Omega^1_{X^{(m)}/\mathbb{C},mx}$ to a regular 1-form on B. Thus we will have, for $\omega \in \Omega^1_{X/\mathbb{C},x}$, an expansion

$$z(t)^*(\operatorname{Tr}\omega) = I(z,\omega)dt + O(t) + (\text{terms not involving } dt), \tag{3.8}$$

where $I(z,\omega)$ is *defined* by the right-hand side of this equation; that is, $I(z,\omega)$ is given by $z(t) \to \operatorname{Tr}\omega\rfloor\partial/\partial t \mod (t)$. We note that where $n = 1$ this definition of $I(z,\omega)$ agrees with that in (2.7) above.

In coordinates, if $z(t)$ is given by (3.1), where the $x_k(t) = \big(\xi_k(t), \eta_k(t)\big)$ are given by Puiseaux series (3.2), then for

$$\omega = f(\xi,\eta)d\xi + g(\xi,\eta)d\eta$$

we will have

$$z(t)^*(\operatorname{Tr}\omega) = \sum_k f(\xi_k,\eta_k)d\xi_k + g(\xi_k,\eta_k)d\eta_k.$$

Expanding f and g in series, using $\sum_{k=0}^{m-1}\epsilon^k = 0$, only integral powers of t survive and, in particular, no negative powers $t^{-\ell/m}$ arising from the $dt^{(n-\ell)/m}$ terms appear. Thus $z(t)^*(\operatorname{Tr}\omega)$ has the form (3.8).

The pairing (3.7) is nondegenerate in the second factor, for the following reason. First, observe another nice general property of traces: for any smooth subvariety $Y \subset X$, traces are *natural for inclusions*, in the sense that for $Y^{(m)} \subset X^{(m)}$ we have

$$\operatorname{Tr}\omega\big|_{Y^{(m)}} = \operatorname{Tr}\big(\omega\big|_Y\big).$$

Second, given any nonzero $\omega \in \Omega^1_{X/\mathbb{C},x}$ we may find a smooth algebraic curve Y passing through x such that $\omega\big|_Y$ is nonzero. We know that the pairing

$$Z^1_{\{x\}}(Y) \otimes_{\mathbb{C}} \Omega^1_{Y/\mathbb{C},x} \to \mathbb{C}$$

is nondegenerate in the second factor, and our claim follows.

The kernel of the surjective mapping

$$Z^n_{\{x\}}(X) \to \operatorname{Hom}^c_{\mathbb{C}}(\Omega^1_{X/\mathbb{C},x}, \mathbb{C})$$

defines an equivalence relation $\sim$ on $Z^n_{\{x\}}(X)$, and our first guess was that the tangent space $T_{\{x\}}Z^n(X)$ should be defined by

$$T_{\{x\}}Z^n(X) = Z^n_{\{x\}}(X)/\sim.$$

As will be explained below, this turned out to be incorrect. In fact, in some sense one knows it *must* be wrong because of the property (3.5). However, $\sim$ is interesting and for later use it is instructive to get some feeling for what it is.

When $n = 2$, working in local coordinates as above we suppose that

$$z(t) = x_1(t) + \cdots + x_m(t)$$

where each

$$x_j(t) = (a_j t, b_j t) + O(t^2) \cdots .$$

The condition that $z(t)$ be $\sim$ equivalent to zero is

$$\begin{cases} \sum a_j = 0 \\ \sum b_j = 0; \end{cases}$$

i.e.

$$\sum_j (a_j, b_j) = 0$$

which is the usual tangent vector property.

Next suppose that

$$\begin{aligned} z(t) &= x_1(t) + \cdots + x_m(t) \\ x_j(t) &= x_j^+(t) + x_j^-(t) \\ x_j^+(t) &= (a_{j_1} t^{1/2} + a_{j_2} t, b_{j_1} t^{1/2} + b_{j_2} t) + O(t^{3/2}) \\ x_j^-(t) &= (-a_{j_1} t^{1/2} + a_{j_2} t, -b_{j_1} t^{1/2} + b_{j_2} t) + O(t^{3/2}). \end{aligned}$$

Now

$$\begin{aligned} I\big(z(t), d\xi\big) &= 2 \sum_j a_{j_2} \\ I(z(t), d\eta) &= 2 \sum_j b_{j_2} \\ I(z(t), \xi d\xi\big) &= \sum_j a_{j_1}^2 \\ I\big(z(t), \eta d\xi\big) &= \sum_j a_{j_1} b_{j_1} \\ I\big(z(t), \xi d\eta\big) &= \sum_j a_{j_1} b_{j_1} \\ I\big(z(t), \eta d\eta\big) &= \sum_j b_{j_1}^2 \\ I\big(z(t), \omega\big) &= 0 \text{ if } \omega \in \mathfrak{m}_x^2 \Omega^1_{X/\mathbb{C},x}. \end{aligned}$$

Then the condition that $z(t)$ be $\sim$ equivalent to zero is expressed by

$$\begin{cases} \sum_j a_{j_1}^2 = 0, \ \sum_j a_{j_2} = 0 \\ \sum_j a_{j_1} b_{j_1} = 0 \\ \sum_j b_{j_1}^2 = 0, \ \sum_j b_{j_2} = 0. \end{cases}$$

In particular, for an irreducible $t^{1/2}$ Puiseaux series

$$\begin{aligned} z(t) = &(a_1 t^{1/2} + a_2 t + \cdots, b_1 t^{1/2} + b_2 t + \cdots) \\ &+(-a_1 t^{1/2} + a_2 t + \cdots, -b_1 t^{1/2} + b_2 t + \cdots) \end{aligned}$$

the information in the equivalence relation $\sim$ is

$$a_1^2, a_1 b_1, b_1^2, a_2, b_2.$$

Thus, b_1/a_1 is determined and to first order $z(t)$ satisfies the equations

$$\begin{aligned} &b_1\xi - a_1\eta + (a_1 b_2 - a_2 b_1)t = 0 \\ &\xi^2 - 2a_2 t\xi - a_1^2 t = 0. \end{aligned}$$

We now turn to the question:

Why should higher degree forms be relevant to tangents to arcs in the space of 0-cycles? One answer stems from Elie Cartan, who taught us that when there are natural parameters in a geometric structure, they should be included as *part* of that structure. Moreover, any use of infinitesimal analysis should include the infinitesimal analysis applied to the parameters as well.[2] In the situation of 0-cycles, there is an infinite dimensional space of parameters—namely, those given in coordinates by the coefficients of all possible Puiseaux series. More interesting in the algebraic case, the spread constructions will give us finite dimensional parameter spaces in which all algebraic relations are preserved.

To systematize this, for a general smooth variety X in terms of local uniformizing parameters around x we denote by P_m the (infinite dimensional) space of coefficients of all formal Puiseaux series as given above for $x_i(t)$. There is a formal map

$$z : \Delta \times P_m \dashrightarrow X^{(m)}$$

given by these Puiseaux series. For any holomorphic q-form Φ defined on $X^{(m)}$ we define the holomorphic $(q-1)$-form $\widetilde{I}(z, \Phi)$ on P_m by

$$z^*(\Phi) = \widetilde{I}(z, \Phi) \wedge dt + O(t) + (\text{terms not involving } dt).^3 \tag{3.9}$$

As we have seen above, $\widetilde{I}(x, \Phi)$ only involves the coefficients of $t^{k/m}$ for $k \leqq m$, and thus it is well defined and holomorphic on a finite dimensional subspace of P_m. In particular we need not worry about issues of convergence. We will see below how $\widetilde{I}(z, \Phi)$ changes when we change the local uniformizing parameters.

Now this construction will make sense when $\dim X = n$ is arbitrary. However, when $n = 1$ so that $X^{(m)}$ is smooth, the 1-forms will generate the holomorphic q-forms as an exterior algebra over $\mathcal{O}_{X^{(m)}}$. This will mean that there is no new intrinsic information in the $\widetilde{I}(z, \Phi)$'s beyond that in the $I(z, \omega)$'s for $\omega \in \Omega^1_{X/\mathbb{C},x}$. As observed in chapter 8 below, $I(z, \omega)$ as defined in (3.1) is the same as $\widetilde{I}(z, \operatorname{Tr}\omega)$ as defined in (3.9).

[2]For example, for a Riemannian manifold—or for that matter any G-structure—Cartan used the full frame bundle, and viewed a connection as a means of infinitesimal displacement in the frame bundle. We will see below that the use of absolute differentials and evaluation maps give a structure similar to a connection on the space of arcs in $Z^n(X)$.

[3]For $q = 1$ and $\Phi = \operatorname{Tr}\omega$, where $\omega \in \Omega^1_{X|\mathbb{C}}$, $\widetilde{I}(z, \operatorname{Tr}\omega)$ is the same as $I(z, \omega)$ defined in (3.1). Only for $q \geqq 2$ will there be a refinement of the definition of $\widetilde{I}(z, \Phi)$ to give what will eventually be denoted by $I(z, \Phi)$ without the tilde. Thus, when $q = 1$ we will always drop the tilde.

For example, when X is an algebraic curve if $z_1, \dots, z_N$ are Puiseaux series for which

$$\sum_\nu n_\nu I(z_\nu, \omega) = 0, \qquad n_\nu \in \mathbb{Z},$$

for all $\omega \in \Omega^1_{X/\mathbb{C},x}$, then we have

$$\sum_\nu n_\nu \widetilde{I}(z_\nu, \Phi) = 0$$

for all Φ. However, as will now be explained, this changes when $n \geqq 2$. Thus, when $n \geqq 2$ new geometric information arise when we introduce the differentials of the Puiseaux series coefficients.

We will work this out in coordinates in the case $n = 2$, as this case will clearly exhibit the central geometric ideas. In chapter 5 we shall reformulate the coordinate discussion intrinsically.

Recall that $Z^2_{\{x\}}(X)$ denotes the group of regular arcs $z(t)$ in $Z^2(X)$ such that $\lim |z(t)| = x$, where $|z(t)|$ is the support of $z(t)$. Any such $z(t)$ may be represented in terms of local uniformizing parameters ξ, η as a sum of irreducible Puiseaux series centered at x. We denote by P the space of coefficients of such Puiseaux series. For any $\Phi \in \Omega^q_{X^{(m)}/\mathbb{C},mx}$ we provisionally define an *abelian invariant* $\widetilde{I}(z, \Phi) \in \Omega^{q-1}_{P/\mathbb{C},z}$ by

$$z(t)^*\Phi = \widetilde{I}(z, \Phi) \wedge dt + O(t) + (\text{terms not involving } dt). \tag{3.10}$$

Because of (3.5) above, all of the information in abelian invariants is captured by the abelian invariants when $\Phi = \operatorname{Tr}\varphi$ is the trace of $\varphi \in \Omega^q_{X/\mathbb{C},x}$ and $q = 1, 2$. In this case we set

$$\widetilde{I}(z, \Phi) = \widetilde{I}(z, \varphi)$$

and provisionally refer to $\widetilde{I}(z, \varphi)$ as a *universal abelian invariant.*[4] The term "universal" refers to the hereditary property of traces discussed above.

Below we will show that the information in the universal abelian invariants $\widetilde{I}(z, \omega)$ for $\omega \in \Omega^1_{X/\mathbb{C},x}$ is invariant under reparametrization of arcs and changes of coordinates. Then we will show that, modulo this information, the information in the universal abelian invariants $\widetilde{I}(z, \varphi)$ for $\varphi \in \Omega^2_{X/\mathbb{C},x}$ is also invariant under reparametrization and coordinate changes. It is here that the differentials of the parameters appear, and the fact that *there is new information* via the universal abelian invariants corresponding to 2-forms is a harbinger of the difference between divisors and cycles in higher codimension.

We consider an arc which is a sum of Puiseaux series in t and $t^{1/2}$—from this the general pattern should be clear. Our family of 0-cycles will thus be a sum

$$z(t) = \sum_\lambda x_\lambda(t) + \sum_\nu x^+_\nu(t) + x^-_\nu(t) \tag{3.11}$$

where

$$x_\lambda(t) = (\alpha_{\lambda_1} t + \cdots, \beta_{\lambda_1} t + \cdots)$$

[4]As explained at the end of this chapter, the final definition of universal abelian invariants only works using absolute differentials in the algebraic setting.

and

$$x_\nu^+(t) = (a_{\nu_1} t^{1/2} + a_{\nu_2} t + \cdots , b_{\nu_1} t^{1/2} + b_{\nu_2} t + \cdots)$$
$$x_\nu^-(t) = (-a_{\nu_1} t^{1/2} + a_{\nu_2} t + \cdots , -b_{\nu_1} t^{1/2} + b_{\nu_2} t + \cdots).$$

Since universal abelian invariants are additive in cycles it will suffice to consider them on

$$z_{\alpha,\beta}(t) = (\alpha_1 t + \cdots , \beta_1 t + \cdots)$$

and

$$z_{a,b} = (a_1 t^{1/2} + a_2 t + \cdots , b_1 t^{1/2} + b_2 t + \cdots)$$
$$+ (-a_1 t^{1/2} + a_2 t + \cdots , -b_1 t^{1/2} + b_2 t + \cdots).$$

Then

$$I(z_{\alpha\beta}, d\xi) = \alpha_1$$
$$I(z_{\alpha\beta}, d\eta) = \beta_1$$

and

$$\widetilde{I}(z_{\alpha\beta}, d\xi \wedge d\eta) = 0;$$

all others are also zero.

The situation for $z_{a,b}$ is more interesting. We have

$$I(z_{ab}, d\xi) = 2a_2$$
$$I(z_{ab}, d\eta) = 2b_2$$
$$I(z_{ab}, \xi d\xi) = a_1^2$$
$$I(z_{ab}, \eta d\eta) = b_1^2$$
$$I(z_{ab}, \xi d\eta) = I(z_{ab}, \eta d\xi) = a_1 b_1$$
$$\widetilde{I}(z_{ab}, d\xi \wedge d\eta) = a_1 db_1 - b_1 da_1;$$

all others are zero. Note that the information in $\widetilde{I}(z_{ab}, d\xi \wedge d\eta)$ is *not* a consequence of that in $I(z_{ab}, \omega)$ and its derivatives for $\omega \in \Omega^1_{X/\mathbb{C},x}$. Also, we note that this new information arises from a 2-form whose trace is not generated by traces of 1-forms.

Referring to (3.11) we see that the universal abelian invariants determine the quantities

$$\begin{cases} \sum_\lambda \alpha_{\lambda,1} + \sum_\nu a_{\nu,2} \\ \sum_\lambda \beta_{\lambda,1} + \sum_\nu b_{\nu,2} \\ \sum_\nu a_{\nu,1}^2, \sum_\nu b_{\nu,1}^2 \\ \sum_\nu a_{\nu,1} b_{\nu,1} \end{cases} \tag{3.12}$$

and

$$\sum_\nu a_{\nu,1} db_{\nu,1} - b_{\nu,1} da_{\nu,1}. \tag{3.13}$$

We note several aspects implied by these computations. The first is that if $z(t)$ is an irreducible Puiseaux series in $t^{1/k}$, only the degree $< k$ homogeneous part of a differential shows up in the universal abelian invariants. We saw this phenomenon earlier for 0-cycles on a curve.

The second is that for a Puiseaux series

$$z(t) = x_1(t) + \cdots + x_k(t)$$

where

$$x_\ell(t) = (a_1\epsilon^\ell t^{1/k} + a_2\epsilon^{2\ell}t^{2/k} + \cdots, b_1\epsilon^\ell t^{\ell/k} + b_2\epsilon^{2\ell}t^{2/k} + \cdots), \qquad \epsilon = e^{2\pi\sqrt{-1}/k}$$

only the coefficients of $t^{m/k}$ for $0 \leqq m \leqq k$ appear in the expression for universal abelian invariants.

The third is that (3.12) may be zero but (3.13) nonzero—*the universal abelian invariants arising from 2-forms will detect subtle geometric/arithmetic information.* More precisely, we see exactly that the new information coming from the 2-forms arises from $\mathrm{Tr}(d\xi \wedge d\eta)$, which as we have seen above is not expressible in terms of traces of lower degree forms. Thus the phenomenon just observed is directly a reflection of (3.5).

We now turn to the behavior of the universal abelian invariants under coordinate changes. For the purposes of illustration we shall take $\alpha_{\lambda,1} = \alpha_\lambda$, $\beta_{\lambda,1} = \beta_\lambda$, $a_{\nu,1} = a_\nu$ and $b_{\nu,1} = b_\nu$. The conclusion we shall draw will hold in generality.

The first fact is

(3.14) *Under a scaling $t \to \mu t$, the universal abelian invariants scale by μ^{-1}*

The point is that no terms $d\mu$ appear. The second fact is that

(3.15) *Under a coordinate change*

$$\xi^\# = f(\xi, \eta)$$
$$\eta^\# = g(\xi, \eta)$$

the universal abelian invariants change by

$$(3.16)\qquad \begin{pmatrix} \sum_\nu a_\nu^{\#2} \\ \sum_\nu a_\nu^\# b_\nu^\# \\ \sum_\nu b_\nu^{\#2} \\ \sum_\lambda \alpha_\lambda^\# \\ \sum_\lambda \beta_\lambda^\# \\ \sum_\nu a_\nu^\# db_\nu^\# - b_\nu^\# da_\nu^\# \end{pmatrix} = \begin{pmatrix} & 0 \\ & \vdots \\ A & \vdots \\ & \vdots \\ & 0 \\ B & C \end{pmatrix} \begin{pmatrix} \sum_\nu a_\nu^2 \\ \sum_\nu a_\nu b_\nu \\ \sum_\nu b_\nu^2 \\ \sum_\lambda \alpha_\lambda \\ \sum_\lambda \beta_\lambda \\ \sum_\nu a_\nu db_\nu - b_\nu da_\nu \end{pmatrix}$$

where A, C are square matrices of complex numbers and B is a matrix with entries in $\Omega^1_{P/\mathbb{C}}$.

The proofs of (3.14) and (3.15) are by direct computation. For example, for

$$z(t) = (at^{1/2} + \cdots, bt^{1/2} + \cdots) + (-at^{1/2} + \cdots, -bt^{1/2} + \cdots) + (\alpha t, \beta t)$$

under a coordinate change

$$\begin{cases} \xi^\# = f(\xi, \eta) \\ \eta^\# = g(\xi, \eta) \end{cases}$$

we have

$$\begin{aligned} a^\# &= af_\xi + bf_\eta \\ b^\# &= ag_\xi + bg_\eta \\ \alpha^\# &= \alpha f_\xi + \beta f_\eta + \frac{a^2}{2} f_{\xi\xi} + abf_{\xi\eta} + \frac{b^2}{2} f_{\eta\eta} \\ \beta^\# &= \alpha g_\xi + \beta g_{\xi\eta} + \frac{a^2}{2} g_{\xi\xi} + abg_{\xi\eta} + \frac{b^2}{2} g_{\eta\eta} \end{aligned}$$

where all partial derivatives are evaluated at x. It follows that the matrix in (3.16) is

(3.17)

$$\begin{pmatrix} f_\xi^2 & 2f_\xi f_\eta & f_\eta^2 & 0 & 0 & 0 \\ f_\xi g_\xi & f_\xi g_\eta + f_\eta g_\xi & f_\eta g_\eta & 0 & 0 & 0 \\ g_\xi^2 & 2g_\xi g_\eta & g_\eta^2 & 0 & 0 & 0 \\ \frac{1}{2} f_{\xi\xi} & f_{\xi\eta} & \frac{1}{2} f_{\eta\eta} & f_\xi & f_\eta & 0 \\ \frac{1}{2} g_{\xi\xi} & g_{\xi\eta} & \frac{1}{2} g_{\eta\eta} & g_\xi & g_\eta & 0 \\ f_\xi dg_\xi - g_\xi df_\xi & * & f_\eta dg_\eta - g_\eta df_\eta & 0 & 0 & f_\xi g_\eta - f_\eta g_\xi \end{pmatrix}$$

where $* = f_\xi dg_\eta + f_\eta dg_\xi - g_\eta df_\xi - g_\xi df_\eta$.

We may summarize this discussion as follows:

For any $z = z(t)$ in $Z^n_{\{x\}}(X)$, the universal abelian invariants

(3.18) $$\widetilde{I}(z, \varphi) \in \Omega^{p-1}_{P/\mathbb{C},z}$$

are defined for $\varphi \in \Omega^p_{X/\mathbb{C},x}$. Here, P is the space of coefficients of Puiseaux series in terms of a choice of parameter t and local uniformizing parameters on X. The vanishing of the universal abelian invariants for all φ and all $p \leqq q$ for fixed q with $1 \leqq q \leqq n$ is independent of these choices.

Looking ahead, there is one more step to be taken before we can define the relation of first order equivalence, to be denoted $\equiv_{1^{\text{st}}}$, on $Z^n_{\{x\}}(X)$. Once this has been done, we can then geometrically define the tangent space by

$$T_{\{x\}} Z^n(X) = Z^n_{\{x\}}(X) / \equiv_{1^{\text{st}}} .$$

This step, which will be seen to make good geometric sense only in the algebraic setting, will consist in working with the *absolute differentials* $\Omega^{\bullet}_{X/\mathbb{Q}}$. In this case, for $\varphi \in \Omega^p_{X/\mathbb{Q},x}$ we will have in place of (3.18) that

$$z(t)^*\varphi = \widetilde{I}(z,\varphi)dt + O(t) + \text{ terms not involving } dt,$$

where

$$\widetilde{I}(z,\varphi) \in \Omega^{p-1}_{P/\mathbb{Q},z}.$$

Then, and this is the point, there is an evaluation map

$$\Omega^{p-1}_{P/\mathbb{Q},x} \xrightarrow{\mathrm{ev}_x} \Omega^{p-1}_{\mathbb{C}/\mathbb{Q}},$$

and we may use this map to define the *universal abelian invariants* $I(z,\varphi)$ by

$$I(z,\varphi) = \mathrm{ev}_x\big(\widetilde{I}(z,\varphi)\big) \in \Omega^{p-1}_{\mathbb{C}/\mathbb{Q}}. \tag{3.19}$$

We will then define

$$z \equiv_{1^{\mathrm{st}}} 0$$

by the condition

$$I(z,\varphi) = 0 \text{ for all } \varphi \in \Omega^p_{X/\mathbb{Q},x} \quad \text{and } 1 \leqq p \leqq n. \tag{3.20}$$

Since automatically $I(z,\varphi) = 0$ if $\varphi \in \Omega^p_{\mathbb{C}/\mathbb{Q}} \subset \Omega^p_{X/\mathbb{Q},x}$, we need only have (3.20) for all

$$\varphi \in \Omega^p_{X/\mathbb{Q},x}/\Omega^p_{\mathbb{C}/\mathbb{Q}}.$$

When $n = 1$ we only have the case $p = 1$ to consider and then

$$\Omega^1_{X/\mathbb{Q},x}/\Omega^1_{\mathbb{C}/\mathbb{Q}} \cong \Omega^1_{X/\mathbb{C},x};$$

this is the reason why absolute differentials did not enter in this case. One reason why forms of higher degree enter when $n \geqq 2$ was explained above.

The above discussion reflects the correct thing to do when X is defined over $\mathbb{Q}$. For more general X, a slight modification is necessary and one must use spreads.

In the next chapter we first present, from a complex analyst's perspective, a digression on absolute differentials and their geometric meaning. We then give an informal, geometric discussion of spreads. In the subsequent chapters we return to the more formal definitions and properties of $\underline{\underline{T}}Z^n(X)$.

Chapter Four

Absolute Differentials (I)

4.1 GENERALITIES

Given a commutative ring R and subring S, one defines the *Kähler differentials* of R over S, denoted

$$\Omega^1_{R/S},$$

to be the R-module generated by all symbols of the form

$$adb, \qquad a, b \in R$$

subject to the relations

$$\text{(4.1)} \qquad \begin{cases} d(a+b) = da + db \\ d(ab) = adb + bda \\ ds = 0 \text{ if } s \in S. \end{cases}$$

In this chapter R will be a field k of characteristic zero, a polynomial ring over k, or a local ring $\mathcal{O}$ with residue field k of characteristic zero. From

$$d(a+a) = 2da$$

it follows that $d2 = 0$, and in fact

$$dp = 0 \qquad p \in \mathbb{Z},$$

and then from $d(q^{-1}) = -q^{-2}dq = 0$ we have

$$d(p/q) = 0 \qquad p, q \in \mathbb{Z}.$$

The Kähler differentials $\Omega^1_{R/\mathbb{Q}}$ are called *absolute differentials*. We shall be primarily concerned with the cases

$$\begin{cases} \Omega^1_{k/\mathbb{Q}} \\ \Omega^1_{\mathcal{O}/k}, \ \Omega^1_{\mathcal{O}/\mathbb{Q}}. \end{cases}$$

If k is a finite extension field of $\mathbb{Q}$, say

$$k = \mathbb{Q}(\alpha)$$

where $\alpha \in \mathbb{C}$ satisfies an irreducible equation

$$f(\alpha) = 0,$$

where $f(x) \in \mathbb{Q}[x]$, then taking the absolute differentials of this equation we have

$$f'(\alpha)d\alpha = 0 \Rightarrow d\alpha = 0$$
$$\Rightarrow \Omega^1_{k/\mathbb{Q}} = 0.$$

More generally, if k is a finitely generated field extension of $\mathbb{Q}$, and if there is

$$f(x_1, \dots, x_n) \in \mathbb{Q}[x_1, \dots, x_n]$$

with

$$f(\alpha_1, \dots, \alpha_n) = 0 \qquad \alpha_1, \dots, \alpha_n \in k,$$

then

$$\sum_i f_{x_i}(\alpha_1, \dots, \alpha_n) d\alpha_i = 0.$$

This gives a linear relation on $d\alpha_1, \dots, d\alpha_n$ in $\Omega^1_{k/\mathbb{Q}}$; in fact, one sees by this line of reasoning (cf. [34]) that

$$\dim_k \Omega^1_{k/\mathbb{Q}} = \operatorname{tr} \deg(k/\mathbb{Q}).$$

Similarly,

$$\dim \Omega^1_{\mathbb{C}/\mathbb{Q}} = \infty$$

with basis $d\alpha_1, d\alpha_2, \dots,$ where $\alpha_1, \alpha_2, \dots$ is a transcendence basis for $\mathbb{C}/\mathbb{Q}$.

If X is a variety defined over a field K with $K \supseteq k$, we will use the notation

$$\Omega^1_{X(K)/k}$$

for the sheaf $\mathcal{O}_{X(K)}$-modules with stalks

$$\Omega^1_{X(K)/k,x} = \Omega^1_{\mathcal{O}_{X(K),x}/k}, \qquad x \in X(K)$$

the Kähler differentials of $\mathcal{O}_{X(K),x}$ over k. When X is defined over $\mathbb{C}$, we will generally write

$$\Omega^1_{X/k} = \Omega^1_{X(\mathbb{C})/k}.$$

There is a natural exact sequence

$$0 \to \Omega^1_{K/k} \otimes \mathcal{O}_{X(K)} \to \Omega^1_{X(K)/k} \to \Omega^1_{X(K)/K} \to 0$$

that we will most often use in the form

$$0 \to \Omega^1_{\mathbb{C}/\mathbb{Q}} \otimes \mathcal{O}_X \to \Omega^1_{X/\mathbb{Q}} \to \Omega^1_{X/\mathbb{C}} \to 0. \tag{4.2}$$

In general we set

$$\Omega^q_{R/S} = \wedge^q \Omega^1_{R/S}.$$

This gives rise to a complex $(\Omega^\bullet_{R/S}, d)$, where $d(a\, db_1 \wedge \cdots \wedge db_q) = da \wedge db_1 \wedge \cdots \wedge db_q$. From (4.2) the R-module $\Omega^q_{X/\mathbb{Q}}$ inherits a decreasing filtration

$$F^m \Omega^q_{X/\mathbb{Q}} = \operatorname{image}\left(\Omega^m_{\mathbb{C}/\mathbb{Q}} \otimes \Omega^{q-m}_{X/\mathbb{Q}} \to \Omega^q_{X/\mathbb{Q}}\right);$$

a differential form belongs to $F^m \Omega^q_{X/\mathbb{Q}}$ if it has at least m differentials of constants. The graded pieces are

$$Gr^m \Omega^q_{X/\mathbb{Q}} \cong \Omega^m_{\mathbb{C}/\mathbb{Q}} \otimes \Omega^{q-m}_{X/\mathbb{C}}.$$

In particular, we note that

$$\Omega^q_{X/\mathbb{Q}} \neq 0 \text{ if } q > \dim X.$$

For example, when X is a curve there is an exact sequence

$$0 \to \Omega^2_{\mathbb{C}/\mathbb{Q}} \otimes \mathcal{O}_X \to \Omega^2_{X/\mathbb{Q}} \to \Omega^1_{\mathbb{C}/\mathbb{Q}} \otimes \Omega^1_{X/\mathbb{C}} \to 0,$$

and we may think of $\Omega^2_{X/\mathbb{Q}}$ as consisting of expressions

$$\begin{cases} d\alpha \wedge d\beta, & \alpha, \beta \in \mathbb{C} \\ d\alpha \wedge \omega, & \alpha \in \mathbb{C}, \omega \in \Omega^1_{X/\mathbb{C}} \end{cases}$$

where in the second expression ω is a "geometric" object (see (4.4) below).

If we have a mapping

$$f : Y \to X$$

between algebraic varieties, then for $x \in X$ the pull-back $f^* : \mathcal{O}_{Y,f(x)} \to \mathcal{O}_{X,x}$ gives an induced mapping of Kähler differentials, and in particular we have

$$f^* : \Omega^q_{X/\mathbb{Q}} \to \Omega^q_{Y/\mathbb{Q}}.$$

Taking for Y a closed point x of X, we have

$$\Omega^q_{x/\mathbb{Q}} \cong \Omega^q_{\mathbb{C}/\mathbb{Q}}$$

and the inclusion $x \hookrightarrow X$ induces an *evaluation map*

$$e_x : \Omega^q_{X/\mathbb{Q}} \to \Omega^q_{\mathbb{C}/\mathbb{Q}}.$$

Explicitly, for $f, g \in \mathcal{O}_{X,x}$, and $\alpha \in \mathbb{C}$

$$\begin{cases} e_x(gd\alpha) = g(x)d\alpha, \\ e_x(gdf) = g(x)d(f(x)). \end{cases}$$

If X is defined over a field k and $f \in \mathcal{O}_{X(k),x}$, then

$$e_x(df) = d(f(x))$$

reflects the field of definition of x.

To give another very concrete example, suppose that $f(x, y) \in \mathbb{C}[x, y]$ so that

$$f(x, y) = 0$$

defines a plane curve X. The *on* X we have the relation

$$f_x dx + f_y dy + \bar{d} f = 0 \tag{4.3}$$

where $\bar{d}$ means "apply $d_{\mathbb{C}/\mathbb{Q}}$ to the coefficients of f." If f is defined over $\mathbb{Q}$—or even over $\bar{\mathbb{Q}}$—this is just the usual relation

$$f_x dx + f_y dy = 0,$$

but (4.3) is more complicated and in some ways more interesting if there are transcendentals among the coefficients of f. We note that a consequence of (4.3) is that on X

$$f_x dx \wedge dy = -\bar{d} f \wedge dy.$$

This allows us to explicitly convert $dx \wedge dy$ into an element of $\Omega^1_{\mathbb{C}/\mathbb{Q}} \otimes \Omega^1_{X/\mathbb{C}}$ as in the above discussion for $\Omega^2_{X/\mathbb{Q}}$ for a curve X.

We also note that for $p \in X$

$$e_p(dx \wedge dy) = -\frac{\bar{d} f(p) \wedge d(y(p))}{f_x(p)},$$

or equivalently

$$e_p(dx \wedge dy) = d(x(p)) \wedge d(y(p)).$$

Thus, $e_p(dx \wedge dy) = 0$ is equivalent to p being defined over an extension field of $\mathbb{Q}$ of transcendence degree at most 1.

Side Discussion: *Kähler differentials don't work analytically.*

A basic principle in complex algebraic geometry is Serre's GAGA (cf. [Serre, J.-P., Géométrie algébrique et géométrie analytique, *Ann. Inst. Fourier* **6** (1956), 1–42]), which operationally says that in the study of complex projective varieties one may work either complex algebraically (Zariski topology, algebraic coherent sheaf cohomology, etc.) or complex analytically (analytic topology, analytic coherent sheaf cohomology, etc.) with the same end result.

From the perspective of this paper a central geometric fact is the following: *In the algebraic category, there is a natural isomorphism of $\mathcal{O}_X$ modules*

$$\Omega^1_{X/\mathbb{C}} \cong \mathcal{O}_X(T^*X). \tag{4.4}$$

Here, the left-hand side is the sheaf of Kähler differentials and the right-hand side is the sheaf of regular sections of the cotangent bundle $T^*X \to X$. This identifies the algebraic object $\Omega^1_{X/\mathbb{C}}$ with the geometric object $\mathcal{O}_X(T^*X)$. As we shall explain in a moment, (4.4) is *false* in the analytic category.

The reason this is important for our study of the tangent spaces to the space of cycles is this: As explained in sections 6.1, 6.2, and 7.3 below, elementary *geometric* considerations of what the relations for first order equivalence of arcs in $Z^p(X)$ should be lead directly to the defining relations (4.1) for Kähler differentials. Because of (4.2) and (4.4), in the *algebraic* category these defining relations then relate directly to the geometry of the variety. But this is no longer true in the analytic category, and this has the implication that our discussion of $TZ^p(X)$ only works in the algebraic setting.

Turning to (4.4) and its failure in the analytic setting, the most näive way to understand the Kähler differential

$$d : \mathcal{O}_X \to \Omega^1_{X/\mathbb{Q}} \tag{4.5}$$

for an algebraic variety X is as follows: First, for $f(x_1, \dots, x_n) \in \mathbb{C}[x_1, \dots, x_n]$ we have

$$df = \sum_i f_{x_i} dx_i + \bar{d} f \tag{4.6}$$

as above. The axioms for Kähler differentials enable us to differentiate a polynomial—viewed as a finite power series—term by term thus giving an explicit formula for df. Using (4.6) we may differentiate a rational function f/g by the quotient rule (restricted, of course, to the open set where $g \neq 0$). Finally, any X is locally an algebraic subvariety of $\mathbb{C}^n$ defined by polynomial equations

$$f_\nu(x_1, \dots, x_n) = 0,$$

and (4.5) is defined by considering $\mathcal{O}_{X,x}$ to be given by the restrictions to X of the rational functions $r = f/g$ on $\mathbb{C}^n$ that are regular near x, defining

$$dr = \frac{df}{g} - \frac{f\,dg}{g^2}$$

using (4.6), and imposing the relations

$$\sum_i f_{\nu,x_i} dx_i + \bar{d} f_\nu = 0.$$

Remark: An interesting fact is that the axioms for Kähler differentials extend formally to allow us to at least formally differentiate term by term the local *analytic* power series of an algebraic function, and when this is done the resulting series converges and gives the correct answer for the absolute differential of the function.

For example, suppose that $f(x, y) = \sum_{p,q=0}^{n} a_{p,q}x^p y^q \in \mathbb{C}[x, y]$, where

$$\begin{cases} f(0, b) = 0 \\ f_y(0, b) \neq 0. \end{cases}$$

Then there is a unique convergent series

$$y(x) = \sum_i b_i x^i, \quad b_0 = b \tag{4.7}$$

satisfying

$$f(x, y(x)) = 0.$$

We note that the coefficients

$$b_i \in \overline{\mathbb{Q}(a_{00}, \ldots, a_{nn})},$$

so that the differentials db_i lie in a finite dimensional subspace $\Omega^1_{\mathbb{C}/\mathbb{Q}}$. Thus the expression

$$\sum_i (db_i + (i+1)b_{i+1}dx)\, x^i \tag{4.8}$$

is a well-defined formal series. We claim that:

(4.9) *If (4.7) converges for $|x| < c$, then (4.8) also converges for $|x| < c$ and represents the absolute differential $dy(x)$ in the sense that we have the identity*

$$-\frac{1}{fy}\left(\bar{d}f + f_x dx\right) = \sum_i (db_i + (i+1)b_{i+1}dx)x^i$$

when the left-hand side is expanded in a series using (4.7).

Before giving the proof we note the

(4.10) **Corollary:** *For $a \in \mathbb{C}$ with $|a| < c$ we have*

$$d_{\mathbb{C}/\mathbb{Q}}(y(a)) = \sum_i db_i a^i + \left(\sum_i (i+1)b_{i+1}a^i\right) da.$$

As will be noted below, this is *false* for a general analytic function. In fact, it is the case that

$$\lim_n a_n = a \Rightarrow \lim_n d_{\mathbb{C}/\mathbb{Q}}(a_n) = d_{\mathbb{C}/\mathbb{Q}}a$$

only for very special limits such as in the finite approximations in (4.7).

Proof of (4.9): Suppose first that $f[x, y] \in \mathbb{Q}[x, y]$. Then the desired identity

$$-f_x/f_y = \sum_i (i+1)b_{i+1}x^i$$

follows from the usual analytic result over $\mathbb{C}$.

Suppose next that $f(x_0, \ldots, x_n, y) \in \mathbb{Q}[x_0, \ldots, x_n, y]$ and the proposed identity in (4.9) is extended to read

$$(4.11) \qquad -\frac{1}{f_y}\left(\bar{d}f + \sum_\alpha f_{x_\alpha} dx_a\right) = \sum_I \left\{db_I + (|i_\alpha| + 1)b_{I+\{\alpha\}}dx_\alpha\right\} x^I,$$

where $x^I = x_0^{i_0} \cdots x_n^{i_n}$ and $I + \{\alpha\} = (i_1, \ldots, i_\alpha + 1, \ldots, i_n)$. By our assumption, $\bar{d}f = 0$ and all $db_I = 0$ so that again (4.11) follows from the usual analytic result.

In general, $f(x, y) \in k[x, y]$, where k is finitely generated over $\mathbb{Q}$. We may write

$$k \cong \mathbb{Q}[x_1, \ldots, x_n]/\{g_1, \ldots, g_m\}$$

where each $g_j \in \mathbb{Q}[x_1, \ldots, x_n]$. Setting $x_0 = x$ we may lift $f(x, y)$ to $f(x_0, x_1, \ldots, x_n, y) \in \mathbb{Q}[x_0, \ldots, x_n, y]$, and then the previous argument applies, taking note of the fact that dividing by $g_1, \ldots, g_m$ does not affect the calculation. □

Remark: In the case when k is a number field the above result follows from Eisenstein's characterization of the power series expansions of algebraic functions defined over k.

Now for a complex analytic variety X^{an} and subfield k of $\mathbb{C}$ we may define the sheaf of Kähler differentials, denoted

$$\Omega^1_{\mathcal{O}_{X^{\text{an}}}/k},$$

as before. This is a sheaf of $\mathcal{O}_{X^{\text{an}}}$-modules and as in the algebraic case there is a map

$$\Omega^1_{\mathcal{O}_{X^{\text{an}}}/\mathbb{C}} \to \mathcal{O}_{X^{\text{an}}}(T^*X^{\text{an}}).$$

However, essentially because the axioms for Kähler differentials only allow finite operations, this mapping is not an isomorphism of $\mathcal{O}_{X^{\text{an}}}$ modules. Concretely, for $X^{\text{an}} = \mathbb{C}$ with coordinate z and letting d denote the Kähler differential map

$$\mathcal{O}_{X^{\text{an}}} \to \Omega^1_{\mathcal{O}_{X^{\text{an}}}/\mathbb{C}},$$

we cannot say that

$$de^z = e^z dz.$$

Note that even though term by term differentiation of the series for e^z makes sense, we have that

$$d(e^z) \neq \sum_n d\left(\frac{z^n}{n!}\right) = \sum_n \frac{z^{n-1}}{(n-1)!}dz$$

where the second equality follows from $d(1/n!) = 0$. In fact

$$0 \neq de = \left(d(e^z)\right)\big|_{z=1} \neq \sum \frac{d1}{(n-1)!} = 0,$$

in contrast to Corollary (4.10).

In concluding this section, for later use we will give one observation that results from the above discussion. Namely, suppose that X is an algebraic variety defined over a field k. If $x_1, \dots, x_n \in k(X)$ give local uniformizing parameters in a neighborhood of $x \in X$, then the $dx_i \in \Omega^1_{X/\mathbb{Q},x}$ give generators for $\Omega^1_{X/\mathbb{C},x}$ as an $\mathcal{O}_{X,x}$-module. If $y_1, \dots, y_n \in k(X)$ is another set of local uniformizing parameters, then each $y_i = y_i(x)$ is an algebraic function of the x_j and we have in $\Omega^1_{X/\mathbb{Q},x}$

$$dy_i = \sum_j \frac{\partial y_i(x)}{\partial x_j} dx_j \qquad \bmod \Omega^1_{k/\mathbb{Q}}.$$

We will write this as

$$\begin{pmatrix} dy_1 \\ \vdots \\ dy_n \end{pmatrix} = J \begin{pmatrix} dx_1 \\ \vdots \\ dx_n \end{pmatrix} + \begin{pmatrix} \alpha_1 \\ \vdots \\ \alpha_n \end{pmatrix} \tag{4.12}$$

where J is the Jacobian matrix and the $\alpha_i \in \Omega^1_{\mathbb{C}/\mathbb{Q}} \otimes \mathcal{O}_{X,x}$. We may symbolically write

$$\alpha_i = \bar{d} y_i(x)$$

where the notation means: "Apply $d_{k/\mathbb{Q}}$ to the coefficients in the series expansion of $y_i(x)$".

Turning to q-forms of higher degree, from (4.12) we have for $q = 2$

$$\begin{aligned} dy_i \wedge dy_j \equiv & \sum_{k,l} \frac{\partial(y_i, y_j)}{\partial(x_k, x_l)} dx_k \wedge dx_l \\ & + \sum_k \left(\frac{\partial y_i}{\partial x_k} d_j - \frac{\partial y_j}{\partial x_k} d_i \right) \wedge dx_k \quad \bmod \Omega^2_{k/\mathbb{Q}}. \end{aligned} \tag{4.13}$$

Comparing this with (3.16) and (3.17) above, we see that the second term in (4.13) has exactly the same formal expression as the matrix B. This will be explained below.

4.2 SPREADS

The fundamental concept of a *spread* dates at least to the mathematicians—especially Weil—who were concerned with the foundations of arithmetic geometry. The modern formulation is due especially to Grothendieck and has been used by Deligne and others in a geometric setting. The idea is that whenever one has an algebraic variety X defined over a field $k \supseteq \mathbb{Q}$, one obtains a family X_s of complex varieties parametrized by the different embeddings

$$s : k \hookrightarrow \mathbb{C}.$$

Anything algebro-geometric that one wishes to study—e.g., the configurations of the subvarieties of X that are defined over k—can and should be done for the entire family.

If we have a presentation

$$k = \mathbb{Q}(\alpha_1, \ldots, \alpha_N)$$

let

$$g_\lambda(\alpha_1, \ldots, \alpha_N) = 0$$

generate the relations over $\mathbb{Q}$ satisfied by the α_i. Then

$$g_\lambda(s_1, \ldots, s_N) \in \mathbb{Q}[s_1, \ldots, s_N]$$

and one may take for the parameter space of the spread the variety

$$S = \operatorname{Var}\{g_\lambda\}.$$

Then S is defined over $\mathbb{Q}$ and

$$\mathbb{Q}(S) \cong k.$$

Every $s \in S(\mathbb{C})$ that does not lie in a proper subvariety of S defined over $\mathbb{Q}$ corresponds to a complex embedding

$$k \hookrightarrow \mathbb{C}$$

given by

$$\alpha_i \to s_i,$$

and this is a one-to-one correspondence.[1]

If X is an algebraic variety defined over k, then for each $s \in S(\mathbb{C})$ as above one obtains a complex variety X_s defined over $\mathbb{Q}(s_1, \ldots, s_N)$. In terms of any algebraic construction, X_s and $X_{s'}$ are indistinguishable if s and s' do not lie in any proper subvariety of $S(\mathbb{C})$ defined over $\mathbb{Q}$. However, the transcendental geometry of X_s and $X_{s'}$ as complex manifolds will generally be quite different. The X_s fit together to form a family

$$\begin{array}{c} \mathcal{X} \\ \downarrow \\ S \end{array} \tag{4.14}$$

where $S = S(\mathbb{C})$, that we call the *spread* of X over k.

Given an algebraic cycle Z on X defined over k—i.e., $Z \in Z^p(X(k))$—one obtains a family of cycles $Z_s \in Z^p(X_s)$ that form a cycle $\mathcal{Z}$ on $\mathcal{X}$ that is defined over $\mathbb{Q}$ — i.e., $\mathcal{Z} \in Z^p(\mathcal{X}(\mathbb{Q}))$. For our purposes, everything is local in the sense that we need only consider the points of S that satisfy no further algebraic relations over $\mathbb{Q}$.

Turning matters around, if we begin with a complex variety X, then X will be defined over a field k as above and we obtain a spread family (4.14) together with a reference point $s_0 \in S$ with

$$X = X_{s_0}.$$[2]

[1]Not every $s \in S(\mathbb{C})$ will give an embedding $k \hookrightarrow \mathbb{C}$. For example, if we take $k = \mathbb{Q}(x)$ then $x \to s$ gives an embedding unless $s \in \bar{\mathbb{Q}}$, when it does not.

[2]To be precise, the spread family (4.14) is specified by the complex variety X together with a choice of field $k \subset \mathbb{C}$ over which X is defined.

Such spread families are in general very special among the local moduli spaces (Kuranishi families) of complex varieties.

The "spread philosophy" is that in any algebro-geometric situation there will be some finitely generated field extension of $\mathbb{Q}$ over which everything is defined, and it is geometrically natural to make use of the resulting spread. This is an essential element in our approach to studying the tangent space to algebraic cycles, and geometrically it is one door through which absolute differentials enter the story.

From a very concrete perspective, let X be an algebraic variety given locally in affine $(n+1)$-space by an equation

$$f(x) = \sum_I \alpha_I x^I = 0$$

where $I = (i_1, \dots, i_{n+1})$ and $x^I = x_1^{i_1} \cdots x_{n+1}^{i_{n+1}}$ (the case where X is defined by several equations will be similar). We may take

$$k = \mathbb{Q}(\cdot\cdot, \alpha_I, \cdot\cdot)$$

to be the field generated by coefficients of $f(x)$, and we let

$$g_\nu(\cdot\cdot, \alpha_I, \cdot\cdot) = 0$$

generate the relations defined over $\mathbb{Q}$ satisfied by these coefficients. We set

$$f(x,s) = \sum_I s_I x^I$$

and may think of the spread as given in (x_i, s_I)-space by the equations

$$\begin{cases} f(x,s) = 0 \\ g_\nu(s) \;\; = 0 \end{cases} \tag{4.15}$$

where $x = (x_1, \dots, x_{n+1})$ and $s = (\cdot\cdot, s_I, \cdot\cdot)$.

As noted above, the algebraic properties over k of X depend only on the polynomial relations over $\mathbb{Q}$ among the coefficients of the defining equation of X. For example, suppose that the equations

$$l_\lambda(x) = \sum_i l_{\lambda i} x_i = 0, \qquad \lambda = 1, \dots, n,$$

where $l_{\lambda i} \in k$ define a line $L \subset X$. This condition may be expressed by

$$f(x) = \sum_\nu h_\lambda(x) l_\lambda(x)$$

where $h_\lambda(x) \in k[x_1, \dots, x_{n+1}]$. We may express the $l_{\lambda i}$ and the coefficients of $h_\lambda(x)$ as being rational functions in $\mathbb{Q}(\cdot\cdot, \alpha_I, \cdot\cdot)$. Replacing $\alpha = (\cdot\cdot, \alpha_I, \cdot\cdot)$ by $s = (\cdot\cdot, s_I, \cdot\cdot) \in S$ we obtain a family of lines $L_s \subset X_s$.

Now let X be a smooth complex algebraic variety defined over a field k, and consider the spread (4.14) with distinguished point $s_0 \in S$ where $X = X_{s_0}$. Then there are fundamental natural identifications

$$\Omega^1_{X(k)/\mathbb{Q}} \cong \Omega^1_{\mathcal{X}(\mathbb{Q})/\mathbb{Q}} \tag{4.16a}$$

and

$$\Omega^1_{S(\mathbb{Q})/\mathbb{Q}, s_0} \cong \Omega^1_{k/\mathbb{Q}} \otimes \mathcal{O}_{S(\mathbb{Q}), s_0}. \tag{4.16b}$$

Here, (4.16a) means the following: Given $x \in X(k)$ there is the corresponding point $(x, s_0) \in \mathfrak{X}(\mathbb{Q})$, and we have

$$\mathcal{O}_{X(k),x} \cong \mathcal{O}_{\mathfrak{X}(\mathbb{Q}),(x,s_0)} \tag{4.16c}$$

and

$$\Omega^1_{X(k)/\mathbb{Q},x} \cong \Omega^1_{\mathfrak{X}(\mathbb{Q})/\mathbb{Q},(x,s_0)}, \tag{4.16d}$$

where the left-hand side is considered as an $\mathcal{O}_{X(k),x}$-module, the right-hand side as an $\mathcal{O}_{\mathfrak{X}(\mathbb{Q}),(x,s_0)}$-module and the identification (4.16c) is made. The differentials are all Kähler differentials over the evident local rings and fields of constants. It is via (4.16a) that absolute differentials may be interpreted as geometric objects, specifically as linear functions on tangent vectors to spreads.

This mechanism is quite clear using the above coordinate description of spreads. Let $f(x, \alpha)$ be the $f(x)$ defining the original X; thus $\alpha_I = s_{0_I}$. Then $\mathcal{O}_{X(k),x}$ is the localization at $x \in X(k)$ of the coordinate ring

$$R_X = k[x_1, \ldots, x_{n+1}]/(f(x, \alpha)), \tag{4.17a}$$

and $\Omega^1_{X(k)/k,x}$ is the $\mathcal{O}_{X(k)}$-module generated by the dx_i and $d\alpha_I$, subject to the relations

$$\begin{cases} \sum_i f_{x_i}(x,\alpha)dx_i + \sum_I f_{s_I}(x,\alpha)d\alpha_I = 0 \\ \sum_I g_{\nu,s_I}(x,\alpha)d\alpha_I = 0 \end{cases} \tag{4.17b}$$

obtained by differentiation of (4.15) and setting $s = s_0$. On the other hand, $\mathcal{O}_{\mathfrak{X}(\mathbb{Q}),(x,s_0)}$ is the localization at (x, s_0) of the coordinate ring

$$R_{\mathfrak{X}} = \mathbb{Q}[x_1, \ldots, x_{n+1}, \ldots, s_I, \ldots]/(f(x,s), \ldots g_\nu(s), \ldots), \tag{4.18a}$$

and $\Omega^1_{\mathfrak{X}(\mathbb{Q})/\mathbb{Q},(x,s_0)}$ is the $\mathcal{O}_{\mathfrak{X}(\mathbb{Q}),(x,s_0)}$-module generated by the dx_i and ds_I, subject to the relations

$$\begin{cases} \sum_i f_{x_i}(x,s)dx_i + \sum_I f_{s_I}(x,s)ds_I = 0 \\ \sum_I g_{\nu,s_I}(x,s)ds_I = 0 \end{cases} \tag{4.18b}$$

obtained from (4.15). Comparing (4.17a), (4.17b) with (4.18a), (4.18b) we see that the map

$$\alpha_I \to s_I$$

establishes the identifications (4.16c), (4.16d).

Finally, among local families of smooth varieties centered at X, spreads have a number of special properties, of which we shall single out three. The first refers to the exact sheaf sequence (4.2):

(4.19) *The extension class of (4.2) may be identified with the Kodaira-Spencer class of the spread family (4.14).*

More precisely, as above fixing a field k over which X is defined, we have the analogue

$$0 \to \Omega^1_{k/\mathbb{Q}} \otimes \mathcal{O}_{X(k)} \to \Omega^1_{X(k)/\mathbb{Q}} \to \Omega^1_{X(k)/k} \to 0 \tag{4.20}$$

of (4.2) for $X(k)$. The extension class of this sequence is an element

$$\begin{aligned} \rho \in H^1\left(\mathrm{Hom}\left(\Omega^1_{X(k)/k}, \Omega^1_{k/\mathbb{Q}} \otimes \mathcal{O}_{X(k)}\right)\right) \\ \wr\| \\ \mathrm{Hom}\left(\Omega^1_{k/\mathbb{Q}}, H^1(\Theta_{X(k)})\right) \\ \wr\| \\ \mathrm{Hom}\left(T_{s_0}S, H^1(\Theta_{X(k)})\right) \end{aligned}$$

where the second identification results from the isomorphism (4.16b). Passing to the corresponding complex varieties, the map

$$T_{s_0}S \xrightarrow{\rho} H^1(\Theta_X)$$

is the Kodaira-Spencer map of the spread family (4.14).

The second property has to do with the evaluation mappings

$$\mathrm{ev}_x : \Omega^1_{X/\mathbb{Q},x} \to \Omega^1_{\mathbb{C}/\mathbb{Q}}. \tag{4.21}$$

These maps are $\mathbb{C}$-linear but not $\mathcal{O}_{X,x}$-linear, and satisfy

$$\mathrm{ev}_x(f d\alpha) = f(x) d\alpha$$

for $d\alpha \in \Omega^1_{\mathbb{C}/\mathbb{Q}}$ and $f \in \mathcal{O}_{X,x}$. Thus we have a $\mathbb{C}$-linear splitting of the exact sequence (4.2), which geometrically may be thought of as something like a connection for the spread family (4.14) but which is only defined along the fiber X_{s_0}

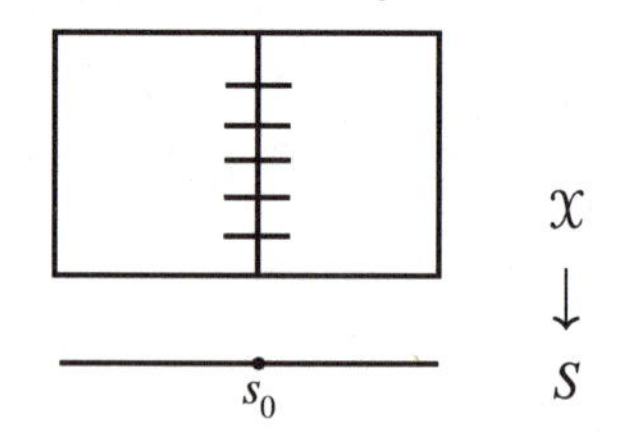

A third property of spreads, one that has been mentioned above, is the following. Let $Z \in Z^p(X(k))$ be an algebraic cycle defined over k. Then the defining ideals for the irreducible components in the support of Z extend naturally from the local rings $\mathcal{O}_{X(k),x}$ to $\mathcal{O}_{\mathbb{Q}(\mathcal{X}),(x,s_0)}$ (cf. (4.16c)). This gives a map

$$Z^p(X(k)) \to Z^p(\mathcal{X}(\mathbb{Q})) \tag{4.22}$$

where in the right-hand side it is understood that we are considering only a Zariski neighborhood of $X = X_{s_0}$ in $\mathcal{X}$. In fact, with this understanding the mapping (4.22) is a bijection. We shall denote by

$$\begin{aligned} \mathcal{Z} \subset \mathcal{X} \\ \downarrow \\ S \end{aligned}$$

the spread of the cycle Z.

Finally, we emphasize again that the spread construction is not unique; we may have different S's for the same k (but always $\mathbb{Q}(S) \simeq k$), and different $\mathcal{Z}$'s for the same Z. As a simple example of the later, suppose that

$$k = \mathbb{Q}(\alpha_1, \dots, \alpha_m)/(g_1, \dots, g_k)$$

and Z is given by

$$p(x_1, \dots, x_n) = 0, \qquad p \in k[x_1, \dots x_n].$$

Then if we replace p by ap, where

$$a = r(\alpha_1, \dots, \alpha_m) \in k,$$

then $\mathcal{Z}$ is changed by $\operatorname{div} r(s_1, \dots, s_m)$ over S.

In [32] we have given a systematic discussion of the ambiguities in the spread construction and of the Hodge-theoretic invariants that remain when one factors them out.

Chapter Five

Geometric Description of $\underline{\underline{T}}Z^n(X)$

5.1 THE DESCRIPTION

Following a standard method in differential geometry, we will describe the tangent space

$$TZ^n_{\{x\}}(X) =: T_{\{x\}}Z^n(X) = Z^n_{\{x\}}(X)/\equiv_{1^{\text{st}}}$$

where $\equiv_{1^{\text{st}}}$ is an equivalence relation given by a subgroup of $Z^n_{\{x\}}(X)$ that we will call *first order equivalence*.

The coordinate description of $\equiv_{1^{\text{st}}}$ is as follows: Denoting by P the space of coefficients of Puiseaux series of arcs in $X^{(m)}$ reducing to mx at $t = 0$, we have a mapping

$$\Omega^q_{X/\mathbb{Q},x} \to \Omega^{q-1}_{P/\mathbb{Q},z}$$

given by

$$\varphi \to \widetilde{I}(z, \varphi) \tag{5.1}$$

where $\widetilde{I}(z, \varphi)$ is the provisional universal abelian invariant associated to an arc

$$z(t) = x_1(t) + \cdots + x_m(t)$$

in $X^{(m)}$ with all $x_i(0) = x$. This map depends on the choice of parameter t and choice of local uniformizing parameters on X used to express the Puiseaux series. We think of P as the union over all m of the coefficients of $t^{k/m}$ for $k \leqq m$. We may compose (5.1) with the evaluation mappings

$$\Omega^{q-1}_{P/\mathbb{Q},z} \to \Omega^{q-1}_{\mathbb{C}/\mathbb{Q}}$$

and extend by linearity in z to obtain a bilinear mapping

$$\Omega^q_{X/\mathbb{Q},x} \otimes_{\mathbb{Z}} Z^n_{\{x\}}(X) \xrightarrow{I} \Omega^{q-1}_{\mathbb{C}/\mathbb{Q}} \tag{5.2}$$

denoted by

$$z \otimes \varphi \to I(z, \varphi).$$

This construction gives our final definition of *universal abelian invariants*. From the explicit expression for these universal abelian invariants it follows that

$$I(z, \alpha \wedge \psi) = \alpha \wedge I(z, \psi) \tag{5.3}$$

for $\alpha \in \Omega^{q-r}_{\mathbb{C}/\mathbb{Q}}$ and $\psi \in \Omega^r_{X/\mathbb{Q},x}$. Taking $r = 0$ we see that

$$I(z, \alpha) = 0 \text{ for } \alpha \in \Omega^q_{\mathbb{C}/\mathbb{Q}},$$

and therefore the pairing (5.2) induces

$$(\Omega^q_{X/\mathbb{Q},x}/\Omega^q_{\mathbb{C}/\mathbb{Q}}) \otimes_{\mathbb{Z}} Z^n_{\{x\}}(X) \to \Omega^{q-1}_{\mathbb{C}/\mathbb{Q}}. \tag{5.4}$$

Proposition: *The pairing (5.4) is nondegenerate on the left; i.e.,*

$$I(z, \varphi) = 0 \text{ for all } z \Rightarrow \varphi \in \Omega^q_{\mathbb{C}/\mathbb{Q}}. \tag{5.5}$$

The proof of this proposition may be given as follows: First, one filters $\Omega^q_{X/\mathbb{Q},x}$ by the images of $\Omega^r_{\mathbb{C}/\mathbb{Q}} \otimes \Omega^{q-r}_{X/\mathbb{Q},x} \to \Omega^q_{X/\mathbb{Q},x}$ and uses (5.3) and induction on q to reduce to proving the result for $\Omega^q_{X/\mathbb{C},x}$. By this we mean that in terms of a fixed set of local uniformizing parameters we use the coordinate formulas as in chapter 3 above, where we map forms in $\Omega^q_{X/\mathbb{C},x}$ to $\Omega^{q-1}_{\mathbb{C}/\mathbb{Q}}$ by the prescriptions given there. Next, we filter $\Omega^q_{X/\mathbb{C},x}$ by $\mathfrak{m}^k_x\Omega^q_{X/\mathbb{C},x}$ with associated graded denoted $Gr^k\Omega^q_{X/\mathbb{C},x}$. Calculations as in chapter 3 then show that

(i) The pairing

$$\begin{pmatrix}\text{Puiseaux series of}\\ \text{order } k+q\end{pmatrix} \otimes Gr^k\Omega^q_{X/\mathbb{C},x} \to \Omega^{q-1}_{\mathbb{C}/\mathbb{Q}}$$

is well-defined, and

(ii) this pairing is nondegenerate on the right.

We illustrate the argument for (ii) when $q = 2$. Pairing a Puiseaux series whose t term is

$$(at^{1/l}, bt^{1/l})$$

against

$$\varphi = \sum A_m \xi^m \eta^{k-m} d\xi \wedge d\eta, \qquad l = k+2$$

gives

$$\sum_m A_m a^m b^{k-m}(adb - bda).$$

Clearly, if this is zero for all a, b then all the A_m vanish and hence $\varphi = 0$. □

Description: *(geometric) We say that an arc $z(t)$ is first order equivalent to zero, written*

$$z(t) \equiv_{1^{\text{st}}} 0,$$

if

$$I(z, \varphi) = 0 \text{ for all } \varphi \in \Omega^q_{X/\mathbb{Q},x} \text{ and } 1 \leqq q \leqq n. \tag{5.6}$$

When the formal definition of $T_{\{x\}}Z^n(X)$ is given in section 7.1 below we will show that

$$T_{\{x\}}Z^n(X) = Z^n_{\{x\}}(X)/\equiv_{1^{\text{st}}}.$$

By the discussion in section 3—cf. (3.18)—together with (5.3), the condition (5.6) is independent of parameter t and local uniformizing parameters on X. In fact, in a moment we shall express $I(z, \varphi)$ in a coordinate-free manner so that this point will be clear.

The reason that the description is said to be "geometric" is that in section 7.1 below we shall reformulate it in terms of $\mathcal{E}xt$'s, and although perhaps less intuitive geometrically this will provide a more satisfactory mathematical definition. In that section, we will give a further justification why the definition takes the form that it does.

Before turning to the intrinsic formulation we want to use (5.3) and the proposition to give an expression for $T_{\{x\}}Z^n(X)$ purely in terms of differentials. For this we have the

Definition: *We define*

$$\mathrm{Hom}^o(\Omega^n_{X/\mathbb{Q},x}, \Omega^{n-1}_{\mathbb{C}/\mathbb{Q}})$$

to be the collections of continuous $\mathbb{C}$*-linear maps*

$$\tau_i : \Omega^i_{X/\mathbb{Q},x} \to \Omega^{i-1}_{\mathbb{C}/\mathbb{Q}}$$

that are $\Omega^\bullet_{\mathbb{C}/\mathbb{Q}}$ *linear in the sense that*

$$\tau_{i+1}(\alpha \wedge \varphi) = \alpha \wedge \tau_i(\varphi)$$

for $\alpha \in \Omega^{n-i}_{\mathbb{C}/\mathbb{Q}}$ *and* $\varphi \in \Omega^i_{X/\mathbb{Q},x}$.

It is easy to see that the τ_i are uniquely determined by τ_n. We set $\tau_n = \tau$ and will think of the elements of $\mathrm{Hom}^o(\Omega^n_{X/\mathbb{Q},x}, \Omega^{n-1}_{\mathbb{C}/\mathbb{Q}})$ as given by $\tau : \Omega^n_{X/\mathbb{Q},x} \to \Omega^{n-1}_{\mathbb{C}/\mathbb{Q}}$ that satisfy the condition that $\tau(\alpha \wedge \varphi) = \alpha$ (something) for $\alpha \in \Omega^{n-i}_{\mathbb{C}/\mathbb{Q}}$. The "something" is then a uniquely determined element $\tau_i(\varphi) \in \Omega^{i-1}_{\mathbb{C}/\mathbb{Q}}$. We obviously have that any such τ is zero on the subspace $\Omega^n_{\mathbb{C}/\mathbb{Q}}$ of $\Omega^n_{X/\mathbb{Q},n}$, so that

$$\mathrm{Hom}^o(\Omega^n_{X/\mathbb{Q},x}, \Omega^{n-1}_{\mathbb{C}/\mathbb{Q}}) \subset \mathrm{Hom}^o_{\mathbb{C}}(\Omega^n_{X/\mathbb{Q},x}/\Omega^n_{\mathbb{C}/\mathbb{Q}}, \Omega^{n-1}_{\mathbb{C}/\mathbb{Q}}).$$

In particular, for $n = 1$

$$\mathrm{Hom}^o(\Omega^1_{X/\mathbb{Q},x}, \mathbb{C}) \cong \mathrm{Hom}^c_{\mathbb{C}}(\Omega^1_{X/\mathbb{C},x}, \mathbb{C}), \tag{5.7}$$

which is the reason that differentials over $\mathbb{Q}$ do not appear in the definition of $T_{\{x\}}Z^n(X)$ when $n = 1$.

From (5.3) and (5.5) we have the

Proposition: *For the stalk at* $x \in X$ *of the tangent sheaf* $\underline{\underline{T}}Z^n(X)$ *described above we have*

$$T_{\{x\}}Z^n(X) \cong \underline{\underline{\mathrm{Hom}}}^o(\Omega^n_{X/\mathbb{Q},x}, \Omega^{n-1}_{\mathbb{C}/\mathbb{Q}})_x. \tag{5.8}$$

In general we note that.

(5.9) *If* $I(z, \psi) = 0$ *for all* $\psi \in \Omega^{p-1}_{X/\mathbb{Q},x}$, *then* $I(z, \varphi)$ *is well-defined for* $\varphi \in \Omega^p_{X/\mathbb{C},x}$.

Thus, we may loosely think of the tangent to an arc $z(t)$ in $Z^n(X)$ to be given by pairing $z(t)$ against all the usual p-forms $\varphi \in \Omega^p_{X/\mathbb{C},x}$ for $1 \leqq p \leqq n$. Here, it is understood that this pairing is given by

$$I(z, \varphi)$$

where $I(z, \varphi)$ is computed in local coordinates as in chapter 3 above, and where for a Puiseaux series coefficient a we understand that

$$da = d_{\mathbb{C}/\mathbb{Q}}a.$$

This is well illustrated by the proof of proposition (5.5) given above.

5.2 INTRINSIC FORMULATION

We shall now formulate the above construction in a coordinate-free manner. Beginning with the case $n = 1$, for a smooth algebraic curve X an arc in $Z^1(X)$ is given by an algebraic curve B with a marked point $t_0 \in B$ together with an algebraic 1-cycle

$$Z \subset B \times X$$

such that for each $t \in B$ the intersection

$$z(t) = Z \cdot (\{t\} \times X)$$

is a 0-cycle on X. Here, we assume that B is smooth, but it need not be complete. To obtain on arc in $Z^1_{\{x\}}(X)$ we assume that

$$|Z| \cdot (\{t_0\} \times X) = \{x\}$$

where $|Z|$ is the support of Z and $\{x\}$ is the support of x (i.e., we ignore multiplicities). With these assumptions we have a natural map

$$\Omega^1_{X/\mathbb{C},x} \to \Omega^1_{B/\mathbb{C},t_0} \tag{5.10}$$

given by

$$\omega \to (\pi_B)_*\big(\pi_X^*\omega\big|_Z\big).$$

If we compose (5.10) with the evaluation map

$$\Omega^1_{B/\mathbb{C},t_0} \to T^*_{t_0}B,$$

then denoting by $Z^1_{\{x\}}(X, B)$ the arcs as above and parametrized by B we have a bilinear pairing

$$\Omega^1_{X/\mathbb{Q},x} \otimes Z^1_{\{x\}}(X, B) \to T^*_{t_0}B. \tag{5.11}$$

Varying B and $Z \subset B \times X$, this pairing is nondegenerate on the left and may be used to define

$$\equiv_{1^{\text{st}}} \subset Z^1_{\{x\}}(X)$$

for curves in a coordinate-free manner.

Of course, for 0-cycles on algebraic curves this construction coincides with the one given in chapter 2. For example, if $z(t)$ is an effective 0-cycle of degree m, then we have a regular mapping

$$f : B \to X^{(m)}$$

with $t_0 \to mx$, and one may verify that

$$(\pi_B)_*(\pi_X^* \omega\big|_Z) = f^*(\mathrm{Tr}\,\omega).$$

Turning to the general case of 0-cycles when $\dim X = n$, we again consider a picture

$$Z \subset B \times X$$

where $\dim B = 1$ and Z is a codimension-n cycle meeting the $\{t\} \times X$'s properly as above and with

$$|Z| \cdot (\{t_0\} \times X) = \{x\}.$$

Beginning with $\omega \in \Omega^n_{X/\mathbb{Q},x}$ we have $\pi_X^* \omega \in \Omega^n_{B\times X/\mathbb{Q},(t_0,x)}$. We may restrict $\pi_X^*\omega$ to Z to obtain

$$(\pi_X^*\omega)\big|_Z =: \omega_Z \in \Omega^n_{Z/\mathbb{Q},(t_0,x)}. \tag{5.12}$$

Then, as discussed in chapter 3, the trace map

$$(\pi_B)_* : \Omega^n_{Z/\mathbb{Q},(t_0,x)} \to \Omega^n_{B/\mathbb{Q},t_0}$$

is defined and

$$(\pi_B)_* \omega_Z =: \omega_B \in \Omega^n_{B/\mathbb{Q},t_0}. \tag{5.13}$$

Since $\dim B = 1$, there is a canonical mapping

$$\Omega^n_{B/\mathbb{Q},t_0} \to \Omega^{n-1}_{\mathbb{C}/\mathbb{Q}} \otimes \Omega^1_{B/\mathbb{C},t_0}. \tag{5.14}$$

Composing (5.12)–(5.14) gives finally

$$\Omega^n_{X/\mathbb{Q},x} \to \Omega^{n-1}_{\mathbb{C}/\mathbb{Q}} \otimes \Omega^1_{B/\mathbb{C},t_0}.$$

This mapping may now be used as in the $n = 1$ case to give a bilinear pairing

$$\Omega^n_{X/\mathbb{Q},x} \otimes Z^n_{\{x\}}(X, B) \xrightarrow{\mathcal{I}_n} \Omega^{n-1}_{\mathbb{C}/\mathbb{Q}} \otimes \Omega^1_{B/\mathbb{C},t_0}$$

whose left kernel, when we vary B and Z, is by proposition (5.5) given by $\Omega^n_{\mathbb{C}/\mathbb{Q}}$.

As in the $n = 1$ case, one may verify that the composition of $\mathcal{I}_n$ with the evaluation map $\Omega^1_{B/\mathbb{C},t_0} \to T^*_{t_0} B$ is the same as the construction via universal abelian invariants given above. The proof depends on the extension to general n of the following lemma, stated here for $n = 2$.

Lemma: *Let X be a smooth algebraic surface and $x, y \in \mathbb{C}(X)$ be rational functions that give local uniformizing parameters in a neighborhood of a point $p \in X$. Let $Y \subset X$ be an algebraic curve with $p \in Y$ and given by*

$$f(x, y) = 0$$

where $f(x, y)$ is an algebraic function of x, y that is regular in a neighborhood of $\big(x(p), y(p)\big)$. Suppose that t is a local uniformizing parameter on Y so that the *map $Y \to X$ is given by $t \to \big(x(t), y(t)\big)$, where $x(t)$, $y(t)$ are algebraic Puiseaux series in t and where $p = \big(x(t_0), y(t_0)\big)$. Then*

$$dx \wedge dy\big|_{f=0} = dx(t) \wedge dy(t),$$

where each side of the equation is interpreted as an element of $\Omega^1_{\mathbb{C}\{t\}/\mathbb{Q},t_0}$. Here, $\mathbb{C}\{t\}$ are the power series of algebraic functions of t that are regular near $t = t_0$.

The proof is by an explicit calculation of the sort done in the preceding section, the point being to use the relation

$$0 = df\big(x(t), y(t)\big) = f_x dx(t) + f_y dy(t) + \bar{d}f$$

where $d = d_{\mathbb{C}\{t\}/\mathbb{Q}}$ and the right hand side is evaluated on $\big(x(t), y(t)\big)$.

We observe that this construction may be used for $\omega \in \Omega^q_{X/\mathbb{Q},x}$ for all $1 \leqq q \leqq n$ to define

$$\Omega^q_{X/\mathbb{Q},x} \otimes Z^n_{\{x\}}(X, B) \xrightarrow{\mathcal{I}_q} \Omega^{q-1}_{\mathbb{C}/\mathbb{Q}} \otimes \Omega^1_{B/\mathbb{C}.t_0}.$$

For $\alpha \in \Omega^p_{\mathbb{C}/\mathbb{Q}}$ and $\varphi \in \Omega^{n-p}_{X/\mathbb{Q},x}$ one sees directly that

$$\mathcal{I}_n\big((\alpha \wedge \varphi) \otimes z\big) = \alpha \wedge \mathcal{I}_{n-p}(\varphi \otimes z)$$

where $\mathcal{I}_{n-p}(\varphi \otimes z) \in \Omega^{n-p-1}_{\mathbb{C}/\mathbb{Q}} \otimes \Omega^1_{B/\mathbb{C},t_0}$ and the notation $\alpha \wedge \mathcal{I}_{n-p}(\varphi \otimes z)$ means to wedge α with the first factor in $\mathcal{I}_{n-p}(\varphi \otimes z)$. This is the intrinsic formulation of the compatibility condition (5.3) above.

This now leads, in the evident way, to the definition of the mapping

$$Z^n_{\{x\}}(X, B) \otimes T_{t_0}B \to \operatorname{Hom}^o(\Omega^n_{X/\mathbb{Q},x}, \Omega^{n-1}_{\mathbb{C}/\mathbb{Q}}). \tag{5.15}$$

Varying B and Z, the mapping (5.15) is surjective and the kernel defines in a coordinate-free manner the relation $\equiv_{1^{\text{st}}}$ of first order equivalence of arcs and tangent space

$$T_{\{x\}}Z^n(X) = Z^n_{\{x\}}(X)/\equiv_{1^{\text{st}}}.$$

We may also use spreads to define the pairing

$$\Omega^n_{X/\mathbb{Q},x} \otimes Z^n_{\{x\}}(X, B) \to \Omega^{n-1}_{\mathbb{C}/\mathbb{Q}} \otimes T^*_{t_0}B. \tag{5.16}$$

If X, z, and B are defined over k, then we have the corresponding spreads $\mathcal{X}$, $\mathcal{B}$ and a cycle $\mathcal{Z}$ on $\mathcal{X} \times_S \mathcal{B}$, together with mappings of differential forms

$$\begin{array}{ccccc} \Omega^n_{\mathcal{X}(\mathbb{Q})/\mathbb{Q}} & \to & \Omega^n_{\mathcal{Z}(\mathbb{Q})/\mathbb{Q}} & \xrightarrow{\mathrm{Tr}} & \Omega^n_{\mathcal{B}(\mathbb{Q})/\mathbb{Q}} \\ & & & & \downarrow \\ & & & & \Omega^{n-1}_{S(\mathbb{Q})/\mathbb{Q}} \otimes \Omega^1_{\mathcal{B}(\mathbb{Q})/S}. \end{array}$$

With the identifications (cf. chapter 4)

$$\begin{cases} \Omega^n_{\mathcal{X}(\mathbb{Q})/\mathbb{Q},(x,s_0)} \cong \Omega^n_{X(k)/\mathbb{Q},x} \\ \Omega^{n-1}_{S(\mathbb{Q})/\mathbb{Q},s_0} \cong \Omega^1_{k/\mathbb{Q}} \\ \Omega^1_{\mathcal{B}(\mathbb{Q})/S} \cong \Omega^1_{\mathcal{B}(k)/k} \end{cases}$$

we obtain a map

$$\Omega^n_{X(k)/\mathbb{Q}} \to \Omega^{n-1}_{k/\mathbb{Q}} \otimes \Omega^1_{B(k)/k}$$

that induces (5.16).

Chapter Six

Absolute Differentials (II)

6.1 ABSOLUTE DIFFERENTIALS ARISE FROM PURELY GEOMETRIC CONSIDERATIONS

In differential geometry the tangent space to a manifold may be defined axiomatically in terms of an equivalence relation on arcs. It would of course be desirable to do the same for the tangent space to the space of cycles. Denoting by $\equiv$ the equivalence we would like to define on an arc $z(t)$ in $Z^n_{\{x\}}(X)$, it should have the following properties:

(i) If $z_1(t) \equiv \tilde{z}_1(t)$ and $z_2(t) \equiv \tilde{z}_2(t)$, then

$$z_1(t) + z_2(t) \equiv \tilde{z}_1(t) + \tilde{z}_2(t);$$

(ii)
$$z(\alpha t) \equiv \alpha z(t) \text{ for } \alpha \in \mathbb{Z}.$$

(iii) If $z(t)$ and $\tilde{z}(t)$ are two arcs in $\mathrm{Hilb}_0(X)$ with the same tangent in $T\,\mathrm{Hilb}_0(X)$, then

$$z(t) \equiv \tilde{z}(t).$$

(iv) If $z(t) = \tilde{z}(t)$ as arcs in $Z^n_{\{x\}}(X)$, then

$$z(t) \equiv \tilde{z}(t).$$

(v) If $\alpha z(t) \equiv \alpha \tilde{z}(t)$ for some nonzero $\alpha \in \mathbb{Z}$, then

$$z(t) \equiv \tilde{z}(t).$$

(vi) If $z(t, u)$ is a 2-parameter family of 0-cycles and we set

$$z_u(t) = z(t, u) \text{ for all } t, u,$$

then if

$$z_u(t) \equiv 0 \text{ for all } u \text{ with } 0 < |u| < \epsilon$$

then

$$z_0(t) \equiv 0.$$

Remark: We do not assume that if there is a sequence $u_i \to 0$ such that

$$z_{u_i}(t) \equiv 0 \text{ for all } i$$

then

$$z_0(t) \equiv 0.$$

Although we do not know that (i)–(vi) give the equivalence relation $\equiv_{1^{st}}$ defined in chapter 5 above, we will show that this is indeed the case in a significant example. Namely, using the notation (f, g) for $\mathrm{Var}(f, g)$ we let

$$z_{\alpha\beta}(t) = (x^2 - \alpha y^2, xy - \beta t), \qquad \alpha \neq 0. \tag{6.1}$$

Then the right-hand side is the arc in the space of 0-cycles in $\mathbb{C}^2$ defined by the ideal generated by $x^2 - \alpha y^2$ and $xy - \beta t$. Denote by F the free group generated by the 0-cycles $z_{\alpha\beta}(t) - z_{1\beta}(t)$ for all $\alpha \in \mathbb{C}^*$, $\beta \in \mathbb{C}$. We will prove the

Proposition: *If $\equiv$ is the minimal equivalence relation satisfying (i)–(vi), then the map $(F/\equiv) \to \Omega^1_{\mathbb{C}/\mathbb{Q}}$ given by*

$$z_{\alpha\beta}(t) \to \beta\frac{d\alpha}{\alpha}$$

is well defined and is an isomorphism.

Thus, the purely geometric axioms (i)–(v) force arithmetic considerations (we will not need (vi)). The proposition has the following

Corollary: *For $\beta \neq 0$, $z_{\alpha\beta}(t) \equiv z_{1,\beta}(t)$ if and only if $\alpha \in \bar{\mathbb{Q}}$.*

Of course, the point is that the geometric condition that $z_{\alpha\beta}(t) \equiv z_{1\beta}(t)$ is equivalent to the arithmetic condition $\alpha \in \bar{\mathbb{Q}}$.

The proof will be given in several steps. Remark that (i) and (iv) will be used in the form

$$(f_1, g) + (f_2, g) \equiv (f_1 f_2, g). \tag{i)$'$}$$

It will be seen that (i)$'$, (ii), and (v) are what force the arithmetic to enter.

The construction

$$z_{\alpha\beta}(t) \mapsto \beta\frac{d\alpha}{\alpha}$$

is a special case of a construction introduced in section 7.1. The properties (i)–(vi) are verified to hold for that general construction.

Step one: We will show that

$$(x^2 - m^2 y^2, xy - t) \equiv (x^2 - y^2, xy - t), \qquad m \in \mathbb{Z}. \tag{6.2}$$

By (i) and (iv),

$$\begin{aligned}(x^2 - m^2 y^2, xy - t) &- (x^2 - y^2, xy - t)\\ &\equiv \left(x^2 - mt, y - \frac{x}{m}\right) + \left(x^2 + mt, y + \frac{x}{m}\right)\\ &\quad -(x^2 - t, y - x) - (x^2 + t, y + x)\end{aligned}$$

(when expanded as sums of Puiseaux series, both sides are the same), and by (iii) the right-hand side is

$$\equiv m\left(x^2 - t, y - \frac{x}{m}\right) + m\left(x^2 + t, y + \frac{x}{m}\right) + (x^2 + t, y - x) + (x^2 - t, y + x) \tag{6.3}$$

which by (i) and (iv) again is

$$\equiv \left(x^2 - t, \left(y - \frac{x}{m}\right)^m (y+x)\right) + \left(x^2 + t, \left(y + \frac{x}{m}\right)^m (y+x)\right).$$

Now using (iii)

$$\left(x^2 - t, \left(y - \frac{x}{m}\right)^m (y+x)\right) \equiv \left(x^2 - t, y^{m+1} + \left(\frac{\binom{m}{2}}{m^2} - 1\right) y^{m-2}x^2 \right.$$
$$\left. + \left(\frac{\binom{m}{3}}{m^3} - \frac{\binom{m}{2}}{m^2}\right) y^{m-3}x^3\right)$$

because $x^4 \equiv t^2 \equiv 0$, and by the same idea the right-hand side is

$$\equiv \left(x^2 - t, y^{m+1} + \left(\left(\frac{\binom{m}{2}}{m^2} - 1\right) y^{m-1} + \left(\frac{\binom{m}{3}}{m^3} - \frac{\binom{m}{2}}{m^2}\right) y^{m-2}x\right) t\right)$$

which by (iii) and (iv) is

$$\equiv (x^2 - t, y^{m-2}) + \left(x^2 - t, y^3 + t\left(\frac{\binom{m}{2}}{m^2} - 1\right) y\right)$$
$$+ \left(x^2 - t, y^3 + t\left(\frac{\binom{m}{3}}{m^3} - \frac{\binom{m}{2}}{m^2}\right) x\right) - (x^2 - t, y^3).$$

Similarly,

$$\left(x^2 + t, \left(y + \frac{x}{m}\right)^m (y-x)\right)$$
$$\equiv (x^2 + t, y^{m-2}) + \left(x^2 + t, y^3 - t\left(\frac{\binom{m}{2}}{m^2} - 1\right) y\right)$$
$$+ \left(x^2 + t, y^3 + t\left(\frac{\binom{m}{3}}{m^3} - \frac{\binom{m}{2}}{m^2}\right) x\right) - (x^2 + t, y^3).$$

Now by (ii)

$$(x^2 - t, y^{m-2}) \equiv -(x^2 + t, y^{m-2})$$
$$(x^2 - t, y^3) \equiv -(x^2 + t, y^3)$$

and

$$\left(x - t, y^3 + t\left(\frac{\binom{m}{2}}{m^2} - 1\right) y\right) \equiv -\left(x + t, y^3 - t\left(\frac{\binom{m}{2}}{m^2} - 1\right) y\right).$$

By (iii) and (iv)

$$\left(x^2 - t, y^3 + t\left(\frac{\binom{m}{3}}{m^3} - \frac{\binom{m}{2}}{m^2}\right) x\right) + \left(x^2 + t, y^3 + t\left(\frac{\binom{m}{3}}{m^3} - \frac{\binom{m}{2}}{m^2}\right) x\right)$$
$$\equiv \left(x^4, y^3 + t\left(\frac{\binom{m}{3}}{m^3} - \frac{\binom{m}{2}}{m^2}\right) x\right)$$

which by (ii) and (iv) is

$$\equiv 4\left(x, y^3 + t\left(\frac{\binom{m}{3}}{m^3} - \frac{\binom{m}{2}}{m^2}x\right)\right)$$

$$\equiv 4(x, y^3) \qquad \text{(by (i))}$$

$$\equiv 0$$

since (x, y^3) is a constant family. Thus all terms in the sum (6.3) cancel out and (6.2) is proved.

Step two: We will prove (6.2) with m^2 replaced by $m \in \mathbb{Z}$. We have

$$\begin{aligned} 2(x^2 - my^2, xy - t) &\equiv \left((x^2 - my^2)^2, xy - t\right) \\ &\equiv \left(x^4 - 2mx^2y^2 + m^2y^4, xy - t\right) \\ &\equiv \left(x^4 - 2mt^2 + m^2y^2, xy - t\right) \\ &\equiv \left(x^4 + m^2y^2, xy - t\right). \end{aligned}$$

Similarly,

$$(x^2 - m^2y^2, xy - t) + (x^2 - y^2, xy - t) \equiv (x^4 + m^2y^2, xy - t)$$

so that

$$2(x^2 - my^2, xy - t) \equiv (x^2 - m^2y^2, xy - t) + (x^2 - y^2, xy - t)$$

which by step one is

$$\equiv 2(x^2 - y^2, xy - t),$$

and then by (v) we get the result.

Step three: We will next establish (6.2) when m^2 is replaced by a rational number $q = m/n$, where m and n are integers. We have by a similar argument to the above

$$\begin{aligned} &\left(x^2 - \left(\frac{m}{n}\right)y^2, xy - t\right) + (x^2 - ny^2, xy - t) \\ &\qquad \equiv (x^2 - my^2, xy - t) + (x^2 - y^2, xy - t). \end{aligned}$$

By step two this gives

$$\left(x^2 - \left(\frac{m}{n}\right)y^2, xy - t\right) + (x^2 - y^2, xy - t) \equiv (x^2 - y^2, xy - t) + (x^2 - y^2, xy - t),$$

which is what was to be proved.

The argument extends immediately to

$$q = q_1 \cdots q_k \tag{6.4}$$

where all $q_i^{n_i} \in \mathbb{Q}$. At this stage we have proved:

If $q \in \bar{\mathbb{Q}}$ is of the form (6.4), then

$$(x^2 - qy^2, xy - t) \equiv (x^2 - y^2, xy - t).$$

Step four: Changing notation slightly, we now set

$$z_{\alpha\beta}(t) = (x^2 - \alpha y^2, xy - \beta t) - (x^2 - y^2, xy - \beta t).$$

By arguments similar to the above we have

$$\begin{aligned}(x^2-\alpha_1y^2, xy-\beta t)&+(x^2-\alpha_2y^2, xy-\beta t)\\ &\equiv (x^2-\alpha_1\alpha_2y^2xy-\beta t)+(x^2-y^2, xy-\beta t)\end{aligned}$$

and

$$(x^2-\alpha y^2, xy-\beta_1t)+(x^2-\alpha y^2, xy-\beta_2t)\equiv(x^2-\alpha y^2, xy-(\beta_1+\beta_2)t).$$

It follows that the map of the free group of cycles of this form modulo the equivalence relation $\equiv$ to $\mathbb{C}^*\otimes_{\mathbb{Z}}\mathbb{C}$ qiven by

$$z_{\alpha,\beta}(t)\to\alpha\otimes\beta$$

is well-defined and surjective. Composing this with the map

$$\mathbb{C}^*\otimes_{\mathbb{Z}}\mathbb{C}\to\Omega^1_{\mathbb{C}/\mathbb{Q}}$$

given by

$$\alpha\otimes\beta\to\beta\frac{d\alpha}{\alpha} \tag{6.5}$$

we obtain the map in the statement of the proposition. To establish that it is injective, we need to know that the defining relations for the Kähler differentials $\Omega^1_{\mathbb{C}/\mathbb{Q}}$ are satisfied by $\equiv$. We thus need to show that the relations

(i) $\alpha_1\alpha_2\otimes\alpha_1\alpha_2-\alpha_1\otimes\alpha_1\alpha_2-\alpha_2\otimes\alpha_1\alpha_2\equiv 0$

(ii) $(\alpha_1+\alpha_2)\otimes(\alpha_1+\alpha_1)-\alpha_1\otimes\alpha_1-\alpha_2\otimes\alpha_2\equiv 0$

(iii) $q\otimes\beta\equiv 0$ if $q\in\mathbb{Q}$

are satisfied. Here, for example, (i) is the Leibniz rule in the form

$$\alpha_1\alpha_2\frac{d(\alpha_1\alpha_2)}{\alpha_1\alpha_2}-\alpha_1\alpha_2\frac{d\alpha_1}{\alpha_1}-\alpha_1\alpha_2\frac{d\alpha_2}{\alpha_2}=0$$

and similarly (ii) is linearity. Now we have just proved (iii), and (i) has been proved above. We thus need to show:

$$\begin{aligned}\big(x^2-(\alpha+\beta)y^2, xy-(\alpha+\beta)t\big)&-\big(x^2-y^2, xy-(\alpha+\beta)t\big)\\ &\equiv(x^2-\alpha y^2, xy-\alpha t)+(x^2-\beta y^2, xy-\beta t)\\ &\quad-(x^2-y^2, xy-\alpha t)-(x^2-y^2, xy-\beta t).\end{aligned}$$

By simple manipulations this is equivalent to

$$\left(x^4+\left(\frac{\alpha+\beta}{\alpha}\right)y^4, xy-\alpha t\right)+\left(x^4+\left(\frac{\alpha+\beta}{\beta}\right)y^4, xy-\beta t\right)\equiv 0. \tag{6.6}$$

Now

$$\begin{aligned}&2\left(x^2-\left(\frac{\alpha+\beta}{\alpha}\right)y^2, xy-\alpha t\right)\\ &\quad\equiv\left(x^2-\left(\frac{\alpha+\beta}{\alpha}\right)^2y^2, xy-\alpha t\right)+(x^2-y^2, xy-\alpha t)\end{aligned}$$

so that

$$\begin{aligned}
2\left[\left(x^2-\left(\frac{\alpha+\beta}{\alpha}\right)y^2, xy-\alpha t\right)+\left(x^2-\left(\frac{\alpha+\beta}{\beta}\right)y^2, xy-\beta t\right)\right]&\\
\equiv\left(x^2-\left(\frac{\alpha+\beta}{\alpha}\right)^2 y^2, xy-\alpha t\right)+(x^2-y^2, xy-\alpha t)&\\
+\left(x^2-\left(\frac{\alpha+\beta}{\beta}\right)^2 y^2, xy-\beta t\right)+(x^2-y^2, xy-\beta t).&
\end{aligned}$$

Now

$$\begin{aligned}
&\left(x^2-\left(\frac{\alpha+\beta}{\alpha}\right)^2 y^2, xy-\alpha t\right)\\
&\quad\equiv\left(x-\left(\frac{\alpha+\beta}{\alpha}\right)y, x^2-(\alpha+\beta)t\right)\\
&\qquad+\left(x+\left(\frac{\alpha+\beta}{\alpha}\right)y, x^2+(\alpha+\beta)t\right)
\end{aligned}$$

and

$$\begin{aligned}
&\left(x^2-\left(\frac{\alpha+\beta}{\beta}\right)^2 y^2, xy-\beta t\right)\\
&\quad\equiv\left(x-\left(\frac{\alpha+\beta}{\beta}\right)y, x^2-(\alpha+\beta)t\right)\\
&\qquad+\left(x+\left(\frac{\alpha+\beta}{\beta}\right)y, x^2+(\alpha+\beta)t\right).
\end{aligned}$$

Thus

$$\begin{aligned}
&2\left[\left(x^2-\left(\frac{\alpha+\beta}{\alpha}\right)y^2, xy-\alpha t\right)+\left(x^2-\left(\frac{\alpha+\beta}{\beta}\right)y^2, xy-\beta t\right)\right]\\
&\equiv\left(x^2-(\alpha+\beta)t, y^2-xy+\left(\frac{\alpha\beta}{\alpha+\beta}\right)y^2\right)\\
&\quad+\left(x^2+(\alpha+\beta)t, y^2+xy+\left(\frac{\alpha\beta}{\alpha+\beta}\right)y^2\right)+2(x^2-y^2, xy-(\alpha+\beta)t)\\
&\equiv\left(x^2-(\alpha+\beta)t, y^2-xy+\left(\frac{\alpha\beta}{\alpha+\beta}\right)t\right)\\
&\quad+\left(x^2+(\alpha+\beta)t, y^2-xy-\left(\frac{\alpha\beta}{\alpha+\beta}\right)t\right)+2(x^2-y^2, xy-(\alpha+\beta)t).
\end{aligned}$$

Now

$$\begin{aligned}
&\left(x^2-(\alpha+\beta)t, y^2-xy+\left(\frac{\alpha\beta}{\alpha+\beta}\right)t\right)\\
&\qquad\equiv-\left(x^2+(\alpha+\beta)t, y^2-xy-\left(\frac{\alpha\beta}{\alpha+\beta}\right)t\right)
\end{aligned}$$

so

$$2\left[\left(x^2-\left(\frac{\alpha+\beta}{\alpha}\right)y^2, xy-\alpha t\right)+\left(x^2-\left(\frac{\alpha+\beta}{\beta}\right)y^2, xy-\beta t\right)\right]$$
$$\equiv 2(x^2-y^2, xy-(\alpha+\beta)t),$$

and (6.6) follows after we cancel the 2's and substitute. □

6.2 A NONCLASSICAL CASE WHEN $n = 1$

In the preceding section we have shown how purely geometric considerations from complex geometry force arithmetic considerations to enter. The example there was for 0-cycles on a surface. However, similar considerations apply also for algebraic curves if we consider the tangent space to "divisors with values in $\mathbb{C}^*$." This will now be explained.

In the remainder of this section, X is a smooth curve. For the sheaf

$$\underline{\underline{Z}}^1(X) = \bigoplus_{x\in X} \underline{\underline{\mathbb{Z}}}_x$$

of divisors on a smooth curve, we have defined the tangent sheaf $\underline{\underline{T}}Z^1(X)$ and shown that it has the description

$$\underline{\underline{T}}Z^1(X) = \bigoplus_{x\in X} \underline{\underline{\mathrm{Hom}}}^c(\Omega^1_{X/\mathbb{C},x}, \mathbb{C}). \tag{6.7}$$

Here, $\underline{\underline{\mathbb{Z}}}_x$ and $\underline{\underline{\mathrm{Hom}}}^c(\Omega^1_{X/\mathbb{C},x}, \mathbb{C})$ are skyscraper sheaves supported at $x \in X$. We now define

$$\underline{\underline{Z}}^1_1(X) = \bigoplus_{x\in X} \underline{\underline{\mathbb{C}}}^*_x.$$

We will show that the natural analogue of (6.7) is

$$\underline{\underline{T}}Z^1_1(X) = \bigoplus_{x\in X} \underline{\underline{\mathrm{Hom}}}^o(\Omega^1_{X/\mathbb{Q},x}, \Omega^1_{\mathbb{C}/\mathbb{Q}}). \tag{6.8}$$

Here, $\mathrm{Hom}^o(\Omega^1_{X/\mathbb{Q},x}, \Omega^1_{\mathbb{C}/\mathbb{Q}})$ are the continuous $\mathbb{C}$-linear homomorphisms

$$\varphi : \Omega^1_{X/\mathbb{Q},x} \to \Omega^1_{\mathbb{C}/\mathbb{Q}}$$

that satisfy

$$\varphi(f\alpha) = \varphi_0(f)\alpha$$

for $f \in \mathcal{O}_{X,x}$ and $\alpha \in \Omega^1_{\mathbb{C}/\mathbb{Q}}$ and where $\varphi_0 \in \mathrm{Hom}^c_{\mathbb{C}}(\mathcal{O}_{X,x}, \mathbb{C})$. The point of (6.8) will be:

(6.9) *The condition that the tangent map*

$$\{\textit{arcs in } Z^1_1(X)\} \to \{\textit{vector space}\}$$

be a homomorphism will imply that, with a natural geometric description of the first order equivalence relation $\equiv_{1^{\mathrm{st}}}$ *on arcs in* $Z^1_1(X)$, *we are led naturally to differentials over* $\mathbb{Q}$.

Thus, in addition to spreads this will give another geometric interpretation of absolute differentials. Still another reason for the appearance of differentials over $\mathbb{Q}$ will be given below when we discuss van der Kallen's description of the formal tangent space to the Milnor K-groups.

The following discussion is heuristic. The intent is to motivate the formal definition. An arc in $Z_1^1(X)$ will be given by $(f(t), g(t))$, where $f(t)$ and $g(t)$ are arcs in $\mathbb{C}(X)^*$, and where it is understood that we consider $g(t)|_{\operatorname{div} f(t)}$ and

$$\operatorname{div} g(t) \cap \operatorname{div} f(t) = \emptyset.$$

Then if $\operatorname{div} f(t) = \sum_i n_i x_i(t)$

$$\bigoplus_i g(t)^{n_i}\Big|_{x_i(t)}$$

will be an arc in $\bigoplus_{x\in X} \underline{\underline{\mathbb{C}}}_x^*$.

To obtain a harbinger of how the geometry and arithmetic will interact, let $\lambda \in \mathbb{C}^*$ be an mth root of unity and let us abbreviate by $\equiv$ the condition that an arc $(f(t), g(t))$ be first order equivalent to zero. Then since the tangent map is a homomorphism

$$(6.10) \qquad m(f(t),\lambda) = \underbrace{(f(t),\lambda) + \cdots + (f(t),\lambda)}_{m \text{ times}} \equiv (f(t),\lambda^m)$$
$$= (f(t), 1)$$
$$= 0$$
$$\Rightarrow m(f(t),\lambda) \equiv 0$$
$$\Rightarrow (f(t),\lambda) \equiv 0 \quad \text{if } \lambda^m = 1,$$

since the stalks of the sheaf $\underline{\underline{TZ}}_1^1(X)$ are vector spaces.

Recall that we have not formally defined $\equiv$; we are just arguing heuristically. The conditions that $\equiv$ should satisfy are

(i) $\big(f(t), g_1(t)\big) + \big(f(t), g_2(t)\big) \equiv \big(f(t), g_1(t)g_2(t)\big)$

(ii) $\big(f_1(t), g(t)\big) + \big(f_2(t), g(t)\big) \equiv \big(f_1(t)\cdot f_2(t), g(t)\big)$

(iii) $\big(f(mt), g(mt)\big) \equiv m\big(f(t), g(t)\big), \qquad m \in \mathbb{Z}$

(iv) $\big(f(t), g(t) + h(t)f(t)\big) \equiv \big(f(t), g(t)\big)$

(v) $\big(f(t), g\big) \equiv 0$ if $\dot{f} \in \mathcal{I}\big(f(0)\big)$

(vi) $(\xi^m + t(g_1+g_2), g_1+g_2) \equiv (\xi^m + tg_1, g_1) + (\xi^m + tg_2, g_2)$.

In (v), the function g is constant in t and $\dot{f} \in \mathcal{I}\big(f(0)\big)$ means that to first order the divisor of f does not move. In (vi), $\xi \in \mathcal{O}_{X,x} \subset \mathbb{C}(X)^*$ is a local uniformizing parameter. As to why (vi) should hold, we have by (iv)

$$(\xi^m + t(g_1+g_2), g_1+g_2) \equiv \left(\xi^m + t(g_1+g_2), -\frac{\xi^m}{t}\right) \qquad (t \neq 0)$$

$$(\xi^m + tg_1, g_1) \equiv \left(\xi^m + tg_1, -\frac{\xi^m}{t}\right) \qquad (t \neq 0)$$

$$(\xi^m + tg_2, g_2) \equiv \left(\xi^m + tg_2, -\frac{\xi^m}{t}\right) \qquad (t \neq 0).$$

Thus by (ii)

$$\begin{aligned}(\xi^m + tg_1, g_1) + (\xi^m + tg_2, g_2) &\\ &\equiv \left((\xi^m + tg_1)(\xi^m + tg_2), -\tfrac{\xi^m}{t}\right) & (t \neq 0)\\ &\equiv \left(\xi^{2m} + t\xi^m(g_1 + g_2), -\tfrac{\xi^m}{t}\right)\\ &\equiv \left(\xi^m, -\tfrac{\xi^m}{t}\right) + \left(\xi^m + t(g_1 + g_2), -\tfrac{\xi^m}{t}\right) & (t \neq 0)\\ &\equiv \left(\xi^m + t(g_1 + g_2), (g_1 + g_2)\right)\end{aligned}$$

by the first step above.

A similar argument shows that

$$\text{(vi)}' \qquad \left(\xi + tf(g_1 + g_2), g_1 + g_2\right) \equiv (\xi + tfg_1, g_1) + (\xi + tfg_2, g_2)$$

where $f \in \mathcal{O}_{X,x} \subset \mathbb{C}(X)$ and where we restrict attention to a Zariski neighborhood of x. From this we infer the

Proposition: *For* $f \in \mathcal{O}_{X,x}$

(a) $(\xi + tf, \lambda) \equiv 0$ *if* $\lambda \in \mathbb{Q}$;

(b) $\left(\xi + tf(g_1 + g_2), g_1 + g_2\right) \equiv (\xi + tfg_1, g_1) + (\xi + tfg_2, g_2)$, *where* $g_1, g_2 \in \mathcal{O}^*_{X,x}$.

Proof: Assertion (b) is (vi)′ above. As for (a) we have

$$\begin{aligned}(\xi + tf, 2) &\equiv \left(\xi + tf\left(\frac{1}{2}(1 + 1)\right), 1 + 1\right)\\ &\equiv \left(\xi + \frac{tf}{2}, 1\right) + \left(\xi + \frac{tf}{2}, 1\right)\\ &\equiv 0.\end{aligned}$$

A similar argument shows that

$$(6.11) \qquad (\xi + tf, m) \equiv 0 \qquad \text{for } m \in \mathbb{N}.$$

Using (6.10) we have

$$(\xi + tf, -1) \equiv 0$$

so that (6.11) holds for all integers m. For $q \neq 0$

$$\begin{aligned}q\left(\xi + tf, \frac{1}{q}\right) &\equiv \underbrace{\left(\xi + tqf\left(\frac{1}{q}\right), \frac{1}{q}\right) + \cdots + \left(\xi + tqf\left(\frac{1}{q}\right), \frac{1}{q}\right)}_{q \text{ times}}\\ &\equiv (\xi + tqf, 1) \qquad \text{(by (vi)}'\text{)}\\ &\equiv 0\\ &\Rightarrow \left(\xi + tf, \frac{1}{q}\right) \equiv 0.\end{aligned}$$

This implies the proposition. □

Using the proposition we see that for $f \in \mathcal{O}_{X,x}$ and $g \in \mathcal{O}^*_{X,x}$

$$(\xi + tf, g) \to f \otimes g \in \mathcal{O}_{X,x} \otimes_{\mathbb{Z}} \mathcal{O}^*_{X,x}$$

is well-defined on arcs modulo the above equivalence relation. It is obviously surjective, and with the obvious notations we have that

$$\begin{cases} f \otimes \lambda \equiv 0 \text{ if } \lambda \in \mathbb{Q}^*, \\ f(g_1 + g_2) \otimes (g_1 + g_2) \equiv fg_1 \otimes g_1 + fg_2 \otimes g_2. \end{cases}$$

From this we conclude that the tangent map from arcs in $Z^1(X, 1)$ factors through the map

$$\mathcal{O}_{X,x} \otimes_{\mathbb{Z}} \mathcal{O}^*_{X,x} \to \Omega^1_{X/\mathbb{Q},x}$$

given by

$$f \otimes g \to f \frac{dg}{g}.$$

In this way differentials over $\mathbb{Q}$ again appear from purely geometric considerations.

Finally we note the following result

(6.12) **Proposition:** *The map to be defined from arcs in* $Z^1_1(X)$ *to* $\underline{\underline{T}} Z^1_1(X)$ *will be surjective.*

We shall not prove this as the much more difficult analogue of this result for the case of surfaces will be given in section 8.2 below. We note that it is a geometric existence result, albeit one that is local (in the Zariski topology) and at the infinitesimal level.

6.3 THE DIFFERENTIAL OF THE TAME SYMBOL

The sheaf $\underline{\underline{Z}}^1_1(X) = \bigoplus_{x \in X} \underline{\underline{\mathbb{C}}}^*_x$ arises by localizing the construction in the Weil reciprocity law, as follows: For $f, g \in \mathbb{C}(X)^*$ and a point $x \in X$, recall the *tame symbol* given by

$$T_x(f, g) = \pm \left(\frac{f^{v_g(x)}}{g^{v_f(x)}} \right)(x) \tag{6.13}$$

where the sign is

$$(-1)^{v_f(x) v_g(x)}.$$

The *Weil reciprocity law* states

$$\prod_{x \in X} T_x(f, g) = 1. \tag{6.14}$$

It is of interest to compute the differential of the map (6.13) and, as an application, determine the infinitesimal form of (6.14).

We begin with a brief digression on residues.

If X is a smooth curve and $x \in X$, then denoting by $\Omega^q_{X/\mathbb{Q}}(nx)$ the absolute q-forms with poles of order at most n at x there are residue maps

$$\begin{aligned} \Omega^1_{X/\mathbb{Q},x}(nx) &\xrightarrow{\mathrm{Res}_x} \mathbb{C} \\ \Omega^2_{X/\mathbb{Q},x}(nx) &\xrightarrow{\mathrm{Res}_x} \Omega^1_{\mathbb{C}/\mathbb{Q}} \\ \Omega^3_{X/\mathbb{Q},x}(nx) &\xrightarrow{\mathrm{Res}_x} \Omega^2_{\mathbb{C}/\mathbb{Q}}, \\ &\vdots \end{aligned}$$

and the residue theorem

$$\sum_{x\in X} \mathrm{Res}_x(\omega) = 0, \qquad \omega \in H^0(\Omega^{q+1}_{X/\mathbb{Q}})$$

holds.

The residues are defined by the sequences

$$\begin{aligned} \Omega^1_{X/\mathbb{Q},x}(nx) &\to \Omega^1_{X/\mathbb{C},x}(nx) \xrightarrow{\mathrm{Res}} \mathbb{C} \\ \Omega^2_{X/\mathbb{Q},x}(nx) &\to \Omega^1_{X/\mathbb{C},x}(nx) \otimes \Omega^1_{\mathbb{C}/\mathbb{Q}} \xrightarrow{\mathrm{Res}\otimes\text{ identity}} \Omega^1_{\mathbb{C}/\mathbb{Q}} \end{aligned}$$

and so forth.

We now consider an arc $(1+tf, g)$ in $\mathbb{C}(X)^* \times \mathbb{C}(X)^*$, where $f \in \mathbb{C}(X)$ and $g \in \mathbb{C}(X)^*$. Denoting by τ the tangent to this arc we have for the differential of the tame symbol dT_x applied to τ that

$$dT_x(\tau) \in \mathrm{Hom}^o(\Omega^1_{X/\mathbb{Q},x}, \Omega^1_{\mathbb{C}/\mathbb{Q}}).$$

Proposition: *For* $\varphi \in \Omega^1_{X/\mathbb{Q},x}$,

$$dT_x(\tau)(\varphi) = \mathrm{Res}_x\left(f\frac{dg}{g} \wedge \varphi\right)$$

when $f dg/g \in \Omega^1_{\mathbb{C}(X)/\mathbb{Q}}$.

Proof: We will check the case where $\nu_f(x) = -1$ and $\nu_g(x) = 0$; the general case is similar. Adjusting constants we may assume that

$$\begin{cases} f = -\frac{1}{\xi}, \\ g = b_0(1 + b_1\xi + \cdots), \qquad b_0 \neq 0 \end{cases}$$

where $\xi \in \mathbb{C}(X)$ is a local uniformizing parameter centered at x. We may also assume that

$$\varphi = c\,d\xi + \alpha + (\text{terms vanishing at } \xi = 0)$$

where $\alpha \in \Omega^1_{\mathbb{C}/\mathbb{Q}}$. Then

$$\mathrm{Res}_x\left(f\frac{dg}{g} \wedge \varphi\right) = -c\frac{db_0}{b_0} + \frac{b_1}{b_0}\alpha \in \Omega^1_{\mathbb{C}/\mathbb{Q}}. \tag{6.15}$$

On the other hand, letting $x(t)$ be the point given by $\xi = t$,

$$T_{x(t)}(1+tf, g)$$

is an arc in $\bigoplus_{x\in X} \underline{\underline{\mathbb{C}}}^*_x$ whose tangent is equal to $dT_x(\tau)$. Now

$$T_{x(t)}(1+tf, g) = (t, b_0(1+b_1t+\cdots)) \in X \times \mathbb{C}^*. \tag{6.16}$$

From the discussion in section 8.2 below we have that the rule for evaluating the tangent to the arc (6.16) on φ is as follows: Denote by u the coordinate on $\mathbb{C}^*$. Then

$$\Phi =: \varphi \wedge \frac{du}{u} \in \Omega^2_{X\times\mathbb{P}^1/\mathbb{Q}}.$$

We restrict this form to the curve B given by the image of (6.16); then

$$\Phi\big|_B \in \Omega^2_{B/\mathbb{Q},t_0}.$$

We then map $\Omega^2_{B/\mathbb{Q},t_0}$ to $\Omega^1_{\mathbb{C}/\mathbb{Q}} \otimes \Omega^1_{B/\mathbb{C},t_0}$ and evaluate the image of $\Phi\big|_B$ on $\partial/\partial t$ in the second factor. Since

$$\Phi\big|_B = (cdt + \alpha + \cdots) \wedge \frac{d(b_0(1+b_1t+\cdots))}{b_0(1+b_1t+\cdots)}$$

we see that this procedure exactly gives (6.15). □

What does the differential dT_x of the tame symbol mean? Above we have taken it to mean the tangent to an arc $T(f(t), g(t))$ in $Z^1_1(X)$, where

$$T = \bigoplus_{x\in X} T_x$$

and $\big(f(t), g(t)\big)$ is an arc in $\mathbb{C}(X)^* \times \mathbb{C}(X)^*$. But the differential should be a linear map

$$dT_x : \text{“?”} \to T_{\{x\}}Z^1_1(X),$$

where $T_{\{x\}}Z^1_1(X)$ is the tangent to arcs $(z(t), a(t))$ in $Z^1_1(X)$ and $z(t)$ is an arc in $Z^1_{\{x\}}(X)$. The issue is what “?” should be?

To get some understanding of this, remark that we could of course take it to be something like

$$T\big(\mathbb{C}(X)^* \times \mathbb{C}(X)^*\big) \cong \mathbb{C}(X) \oplus \mathbb{C}(X).$$

But because of the easily verified relations

$$\begin{cases} T_x(f^n, g) = T_x(f, g^n) & n \in \mathbb{Z}, \\ T_x(f_1f_2, g) = T_x(f_1, g)T_x(f_2, g) \\ T_x(f, g_1g_2) = T_x(f, g_1)T_x(f, g_2) \end{cases} \tag{6.17}$$

we see that the tame symbol really should be defined on

$$\mathbb{C}(X)^* \otimes_{\mathbb{Z}} \mathbb{C}(X)^*.$$

However, in addition to (6.17) one may directly check that

$$T_x(f, 1-f) = 0, \qquad f \in \mathbb{C}(X)^*\backslash\{1\}.$$

At this point we recall the basic

Definition: *For any field F of characteristic zero, or a local ring whose residue field is of characteristic zero, the group $K_2(F)$ is defined by*

$$K_2(F) = F^* \otimes_{\mathbb{Z}} F^*/R$$

where R is the subgroup generated by the Steinberg relation

$$a \otimes (1-a), \qquad a \in F^* \backslash \{1\}.$$

The image of $a \otimes b \in F^* \otimes_{\mathbb{Z}} F^*$ in $K_2(F)$ is denoted by $\{a, b\}$. Our reference for K-theory is the book [35] by Milnor. We have taken as definition the identification given by the theorem of Matsumura. We note that the above definition will extend to define the *higher Milnor K-groups*

$$K_p^M(F) = \overbrace{F^* \otimes \cdots \otimes F^*}^{p}/R, \qquad p \geqq 2,$$

where R is generated by elements $a_1 \otimes \cdots \otimes a_k \otimes (1-a_k) \otimes \cdots \otimes a_{p-1}$ obtained by putting a Steinberg relation in adjacent positions.

Of particular importance for our work is the formal tangent space to $K_2(F)$. Letting ϵ be a formal indeterminate satisfying $\epsilon^2 = 0$ and $F^*[\epsilon]$ the set of expressions

$$\begin{cases} a + \epsilon b \\ a \in F^*, b \in F \end{cases}$$

we observe that we may define $K_2(F^*[\epsilon])$ as above but now with $F^*[\epsilon]$ replacing F^*.

Definition: *The formal tangent space $TK_2(F)$ is defined by*

$$TK_2(F) = \ker\{K_2(F[\epsilon]) \to K_2(F)\}.$$

Here the restriction map is obtained by setting $\epsilon = 0$. A central result is the

Theorem (van der Kallen [12]): *There is a natural identification*

$$TK_2(F) \cong \Omega^1_{F/\mathbb{Q}}. \tag{6.18}$$

The map that induces (6.18) is

$$\{1 + \epsilon a, b\} \to \frac{adb}{b}$$

where $a \in F, b \in F^*$. Because of its central role in what follows, in the appendix to this section we have given a proof of the theorem of van der Kallen. Remark that the proof extends to the higher Milnor K-groups to give a natural identification

$$TK_p^M(F) \cong \Omega^{p-1}_{F/\mathbb{Q}}$$

where the map is

$$\{1 + \epsilon a, b_1, \ldots, b_{p-1}\} \to \frac{adb_1}{b_1} \wedge \cdots \wedge \frac{db_{p-1}}{b_{p-1}}.$$

Returning to the differential of the tame symbol, using (6.18) we should have

$$dT_x : \Omega^1_{\mathbb{C}(X)/\mathbb{Q},x} \to \mathrm{Hom}^o(\Omega^1_{X/\mathbb{Q},x}, \Omega^1_{\mathbb{C},\mathbb{Q}}). \tag{6.19}$$

Comparing (6.14) with the proposition above we infer the

Proposition: *The differential (6.19) of the tame symbol is given by*

$$dT_x(\psi)(\varphi) = \operatorname{Res}_x(\psi \wedge \varphi) \tag{6.20}$$

where $\psi \in \Omega^1_{\mathbb{C}(X)/\mathbb{Q}}$ *and* $\varphi \in \Omega^1_{X/\mathbb{Q},x}$.

We may now turn the discussion around. There is a sheaf mapping

$$\underline{\underline{\Omega}}^1_{\mathbb{C}(X)/\mathbb{Q}} \xrightarrow{\rho} \bigoplus_{x \in X} \underline{\underline{\mathrm{Hom}}}^o(\Omega^1_{X/\mathbb{Q},x}, \Omega^1_{\mathbb{C}/\mathbb{Q}})$$

given on the stalk at a point x by

$$\rho(\psi)(\varphi) = \operatorname{Res}_x(\psi \wedge \varphi) \in \Omega^1_{\mathbb{C}/\mathbb{Q}}.$$

It is easy to check that this mapping is surjective and has kernel $\Omega^1_{X/\mathbb{Q},x}$. Thus we have the exact sheaf sequence

$$0 \to \Omega^1_{X/\mathbb{Q}} \to \underline{\underline{\Omega}}^1_{\mathbb{C}(X)/\mathbb{Q}} \to \bigoplus_{x \in X} \underline{\underline{\mathrm{Hom}}}^o(\Omega^1_{X/\mathbb{Q},x}, \Omega^1_{\mathbb{C}/\mathbb{Q}}) \to 0 \tag{6.21}$$

which gives a flasque resolution of $\Omega^1_{X/\mathbb{Q}}$. There is also the following special case of the Bloch-Gersten-Quillen flasque resolution of the sheaf $\mathcal{K}_2(\mathcal{O}_X)$:

$$0 \to \mathcal{K}_2(\mathcal{O}_X) \to \underline{\underline{K}}_2\big(\mathbb{C}(X)\big) \xrightarrow{T} \bigoplus_{x \in X} \underline{\underline{\mathbb{C}}}^*_x \to 0. \tag{6.22}$$

From the above discussion and proposition we infer that

The tangent sequence to (6.22) may be defined and it is then given by (6.21).

This is the "higher" analogue of the discussion in section 2 which identified

$$0 \to \mathcal{O}_X \to \underline{\underline{\mathbb{C}}}(X) \to \bigoplus_{x \in X} \underline{\underline{\mathrm{Hom}}}^c(\Omega^1_{X/\mathbb{C},x}, \mathbb{C}) \to 0$$

as the tangent sequence of

$$0 \to \mathcal{O}^*_X \to \underline{\underline{\mathbb{C}}}(X)^* \to \bigoplus_{x \in X} \underline{\underline{Z}}_x \to 0.$$

Turning now to the infinitesimal form of the Weil reciprocity law, we observe that there is a mapping of sheaves

$$\bigoplus_{x \in X} \mathbb{C}^*_x \to \underline{\underline{\mathbb{C}}}^* \tag{6.23}$$

given by

$$\bigoplus_{x \in X} (x, \lambda_x) \to \prod_{x \in X} \lambda_x$$

where $\underline{\underline{\mathbb{C}}}^*$ is the constant Zariski sheaf with stalks $\underline{\underline{\mathbb{C}}}^*_x = \mathbb{C}^*$. The differential of (6.23) is a mapping of sheaves

$$\bigoplus_{x \in X} \underline{\underline{\mathrm{Hom}}}^o(\Omega^1_{X/\mathbb{Q},x}, \Omega^1_{\mathbb{C}/\mathbb{Q}}) \to \underline{\underline{\mathbb{C}}} \tag{6.24}$$

that may be described as follows: An element τ_x of $\mathrm{Hom}^o(\Omega^1_{X/\mathbb{Q},x}, \Omega^1_{\mathbb{C}/\mathbb{Q}})$ satisfies

$$\tau_x(h\alpha) = \tau_x^o(h)\alpha$$

for $\alpha \in \Omega^1_{\mathbb{C}/\mathbb{Q}}$, $h \in \mathcal{O}_{X,x}$ and where $\tau_x^o \in \mathrm{Hom}^c(\mathcal{O}_{X,x}, \mathbb{C})$. Then the differential of (6.23) at $\bigoplus_{x\in X}(x, \lambda_x)$ is given by sending $\tau = \bigoplus_{x\in X}\tau_x$, where $\tau_x \in \mathrm{Hom}^o(\Omega^1_{X/\mathbb{Q},x}, \Omega^1_{\mathbb{C}/\mathbb{Q}})$, to $\sum_{x\in X}\tau_x^o(1)$; i.e., we have

$$\tau \to \sum_{x\in X}\tau_x^o(1). \tag{6.25}$$

Next, as noted above the differential of the tame symbol applied to the image of the arc $\{1 + tf, g\}$ in $K_2(\mathbb{C}(X))$ is the tangent vector $\tau = \{\tau_x\}_{x\in X} \in \bigoplus_{x\in X}\mathrm{Hom}^o(\Omega^1_{X/\mathbb{Q},x}, \Omega^1_{\mathbb{C}/\mathbb{Q}})$ given by

$$\tau_x(\varphi) = \mathrm{Res}_x\left(f\frac{dg}{g}\wedge\varphi\right), \qquad \varphi \in \Omega^1_{X/\mathbb{Q},x}.$$

Taking $\varphi = h\alpha$ where $h \in \mathcal{O}_{X,x}$ we have

$$\tau_x(h\alpha) = \mathrm{Res}_x\left(hf\frac{dg}{g}\right)\alpha;$$

i.e.

$$\tau_x^o(h) = \mathrm{Res}_x\left(hf\frac{dg}{g}\right).$$

Referring to (6.25) and taking $h = 1$, we have the

Proposition: *The infinitesimal form of the Weil reciprocity law is given by the residue theorem*

$$\sum_{x\in X}\mathrm{Res}_x\left(f\frac{dg}{g}\right) = 0.$$

The Weil reciprocity law is the first of a series of reciprocity laws whose higher versions are due to Suslin. To state them above we have defined the higher Milnor K-groups by

$$K_p^M(F) = \overbrace{F^* \otimes_{\mathbb{Z}} \cdots \otimes_{\mathbb{Z}} F^*}^{p}/R$$

where R is generated by putting $a \otimes (1-a)$ $(a \in F^*\backslash\{1\})$ in adjacent positions. As noted there, the van der Kallen theorem generalizes in a straightforward way to give for the formal tangent space

$$TK_p^M(F) \cong \Omega^{p-1}_{F/\mathbb{Q}}, \qquad p \geqq 1. \tag{6.26}$$

Denoting the element of $K_p^M(F)$ that is the image of $a_1 \otimes \cdots \otimes a_p$ by $\{a_1, \ldots, a_p\}$, the map that gives (6.26) is

$$\{1 + \epsilon a, b_1, \ldots, b_{p-1}\} \to a\frac{db_1}{b_1}\wedge\cdots\wedge\frac{db_{p-1}}{b_{p-1}}$$

where $a \in F$ and $b_1, \ldots, b_{p-1} \in F^*$.

Returning to the consideration of an algebraic curve X, there is for each $x \in X$ a map

$$K_p^M(\mathbb{C}(X)) \xrightarrow{\partial_x} K_{p-1}(\mathbb{C}_x)$$

given by alternating the tame symbol as in the formula for a coboundary. The *Suslin reciprocity formula* states that the composite

(6.27)
$$\overbrace{K_p^M(\mathbb{C}(X)) \xrightarrow{\partial_x} \bigoplus_{x\in X} K_{p-1}^M(\mathbb{C})_x \to K_{p-1}^M(\mathbb{C})}^{\partial},$$

is the identity, where

$$\partial = \prod_{x\in X} \partial_x.$$

The infinitesimal version of (6.27) states that the composite

(6.28)
$$\overbrace{\Omega^{p-1}_{\mathbb{C}(X)/\mathbb{Q}} \to \bigoplus_{x\in X} \Omega^{p-2}_{\mathbb{C}_x/\mathbb{Q}} \to \Omega^{p-2}_{\mathbb{C}/\mathbb{Q}}}^{d(\partial)}$$

is zero, where $d(\partial)$ is the differential of ∂ using the identification (6.26). One may check that for $\varphi \in \Omega^{p-1}_{\mathbb{C}(X)/\mathbb{Q}}$

$$d(\partial)(\varphi) = \sum_{x\in X} \mathrm{Res}_x(\varphi).$$

Then as in the $p = 2$ case of Weil reciprocity, we have the

Proposition: *The infinitesimal form of the Suslin reciprocity law is given by the residue theorem*

$$\sum_{x\in X} \mathrm{Res}_x(\varphi) = 0 \quad \text{in } \Omega^{p-2}_{\mathbb{C}/\mathbb{Q}}.$$

The higher reciprocity theorems are in some ways more subtle than Weil reciprocity. For example, when $p = 3$, since

$$d(\partial_x)\left(f\frac{dg_1}{g_1} \wedge \frac{dg_2}{g_2}\right) = \mathrm{Res}_x\left(f\frac{dg_1}{g_1} \wedge \frac{dg_2}{g_2}\right) \in \Omega^1_{\mathbb{C}/\mathbb{Q}},$$

we see that

$$\sum_{x\in X} d(\partial_x)\left(f\frac{dg_1}{g_1} \wedge \frac{dg_2}{g_2}\right) = 0$$

gives an *arithmetic* relation in the values of g_1, g_2 at the points of $|\mathrm{div}\, f| \cup |\mathrm{div}\, g_1| \cup |\mathrm{div}\, g_2|$. As we will see below, this is very much the flavor of what happens for rational equivalence in higher codimension.

6.3.1 Appendix: On the theorem of van der Kallen

We want to discuss the natural isomorphism

$$TK_2(F) \cong \Omega^1_{F/\mathbb{Q}} \tag{6.29}$$

due to van der Kallen [12]. Here, F is either a field of characteristic zero or a local ring with residue field of characteristic zero. Following that we will discuss how *geometric* considerations inevitably lead into $K_2(\mathcal{O}_X)$.

We begin by recalling that

$$K_2(F) = F^* \otimes_{\mathbb{Z}} F^*/R$$

where R is generated by the Steinberg relations

$$a \otimes (1-a) \sim 1, \qquad a \in F^* \backslash \{1\}. \tag{6.30}$$

The following are nontrivial but elementary consequences of the defining relations for $\otimes_{\mathbb{Z}}$ and (6.30) (cf. Milnor [35])

$$\begin{cases} \{a, 1\} = 1 \\ \{a, b\} = \{b, a\}^{-1} \\ \{a_1 a_2, b\} = \{a_1, b\}\{a_2, b\} \\ \{a, b_1 b_2\} = \{a, b_1\}\{a, b_2\} \\ \{a, -a\} = 1 \end{cases} \tag{6.31}$$

Recall also that by definition

$$TK_2(F) = \ker\{K_2(F[\epsilon]) \xrightarrow{\epsilon=0} K_2(F)\}.$$

Lemma: *$TK_2(F)$ is generated by elements*

$$\{1 + \epsilon a, b\}, \qquad a \in F, b \in F^*. \tag{6.32}$$

Assuming for a moment the lemma, the mapping (6.30) is induced by

$$\{1 + \epsilon a, b\} \to a\frac{db}{b}. \tag{6.33}$$

Since (6.33) is clearly surjective, we have to show that it is well-defined and injective.

Proof of lemma: Suppose that $\{a_1 + \epsilon b_1, a_2 + \epsilon b_2\} \in K_2(F[\epsilon])$ and $\{a_1, a_2\} = 1$. Then by (6.31)

$$\begin{aligned} \{a_1 + \epsilon b_1, a_2 + \epsilon b_2\} &= \{a_1, a_2 + \epsilon b_2\}\left\{1 + \epsilon\frac{b_1}{a_1}, a_2 + \epsilon b_2\right\} \\ &= \{a_1, a_2\}\left\{a_1, 1 + \epsilon\frac{b_2}{a_2}\right\}\left\{1 + \epsilon\frac{b_1}{a_1}, a_2 + \epsilon b_2\right\} \\ &= \left\{a_1, 1 + \epsilon\frac{b_2}{a_2}\right\}\left\{1 + \epsilon\frac{b_1}{a_1}, a_2\right\}\left\{1 + \epsilon\frac{b_1}{a_1}, 1 + \epsilon\frac{b_2}{a_2}\right\}. \end{aligned}$$

We are reduced to proving: For $a, b \neq 0$

(6.34) $\{1+\epsilon a, 1+\epsilon b\}$ is a product of elements of the form $\{1+\epsilon f, g\}$ where $f, g \in F^*$.

For $c \neq 0$,

$$\begin{aligned}\{1+\epsilon a, 1+\epsilon b\} &= \left\{1+\epsilon a, \frac{1}{c}\right\}\{1+\epsilon a, c+\epsilon bc\}\\ &= \left\{1+\epsilon a, \frac{1}{c}\right\}\{1+\epsilon a, 1-(1-c-\epsilon bc)\}\\ &= \left\{1+\epsilon a, \frac{1}{c}\right\}\{(1+\epsilon a)(1-c-\epsilon bc), 1-(1-c-abc)\}\end{aligned}$$

by using bilinearity and the Steinberg relation

$$= \left\{1+\epsilon a, \frac{1}{c}\right\}\{1-c+\epsilon\big(a(1-c)-bc\big), c+\epsilon bc\}.$$

Setting $c = a/(a+b)$ under the assumption $a+b \neq 0$, the term $a(1-c)-bc$ drops out and we have

$$\begin{aligned}&= \left\{1+\epsilon a, \frac{a+b}{a}\right\}\{1-c, c+\epsilon bc\}\\ &= \left\{1+\epsilon a, \frac{a+b}{a}\right\}\{1-c, 1+\epsilon b\}\\ &= \left\{1+\epsilon a, \frac{a+b}{a}\right\}\left\{\frac{-b}{a+b}, 1+\epsilon b\right\}\\ &= \{1+\epsilon a, a+b\}\{1+\epsilon a, b\}^{-1}\{-b, 1+\epsilon b\}\{1+\epsilon b, a+b\}\\ &= \{1+\epsilon(a+b), a+b\}\{1+\epsilon a, a\}^{-1}\{1+\epsilon b, b\}^{-1}\{-1, 1+\epsilon b\},\end{aligned}$$

which using $\{-1, 1+\epsilon b\} = \{1, 1+\epsilon b/2\} = 1$ gives

$$= \{1+\epsilon(a+b), a+b\}\{1+\epsilon a, a\}^{-1}\{1+\epsilon b, b\}^{-1}.$$

This proves (6.34) and therefore the lemma. □

We note that under the mapping (6.33)

$$\{1+\epsilon(a+b), a+b\}\{1+\epsilon a, a\}^{-1}\{1+\epsilon b, b\}^{-1} \to 0.$$

Since the relations

$$\{1+\epsilon ka, b\} = \{1+\epsilon a, b^k\}, \qquad k \in \mathbb{Z}$$

obviously also map to zero it follows that:

the mapping (6.28) defines $TK_2(F) \to \Omega^1_{F/\mathbb{Q}}$.

We shall now prove the

Lemma: $\{1 + \epsilon a, 1 + \epsilon b\} = 1$ *in* $K_2(F[\epsilon])$.

Proof: From the proof of the preceding lemma we have

$$\{1 + \epsilon A, 1 + \epsilon B\} = \frac{\{1 + \epsilon(A + B), A + B\}}{\{1 + \epsilon A, A\}\{1 + \epsilon B, B\}} \tag{6.35}$$

for any $A, B \in F^*$ such that also $A + B \in F^*$. Thus

$$\begin{aligned}\{1 + \epsilon A, 1 + \epsilon B\} &= \{(1 + \epsilon(A - B))(1 + \epsilon B), 1 + \epsilon B\} \\ &= \{1 + \epsilon(A - B), 1 + \epsilon B\}\{1 + \epsilon B, 1 + \epsilon B\},\end{aligned}$$

which using

$$\begin{aligned}\{1 + \epsilon B, 1 + \epsilon B\} &= \{1 + \epsilon B, -(1 + \epsilon B)\}\{1 + \epsilon B, -1\} \\ &= \{1 + \epsilon B, -1\} = \{1 + \epsilon B/2, -1\}^2 = \{1 + \epsilon B/2, 1\} = 1\end{aligned}$$

gives $\{1 + \epsilon A, 1 + \epsilon B\} = \{1 + \epsilon(A - B), 1 + \epsilon B\}$.

Applying (6.35) to both sides of the equation and assuming that $A - B \in F^*$ gives

$$\frac{\{1 + \epsilon A, A\}}{\{1 + \epsilon(A - B), A - B\}\{1 + \epsilon B, B\}} = \frac{\{1 + \epsilon(A + B), A + B\}}{\{1 + \epsilon A, A\}\{1 + \epsilon B, B\}}$$

$$\Rightarrow \{1 + \epsilon A, A\}^2 = \{1 + \epsilon(A + B), A + B\}\{1 + \epsilon(A - B), A - B\}.$$

Setting $u = A + B$ and $v = A - B$ this becomes

$$\left\{1 + \epsilon\frac{(u + v)}{2}, \frac{u + v}{2}\right\}^2 = \{1 + \epsilon u, u\}\{1 + \epsilon v, v\}.$$

The left-hand side is

$$\left\{1 + \epsilon(u + v), \frac{u + v}{2}\right\} = \left\{1 + \epsilon(u + v), u + v\right\}\left\{1 + \epsilon(u + v), \frac{1}{2}\right\}.$$

By (iv) below the last term on the right is equal to 1, so that we obtain

$$\{1 + \epsilon(u + v), u + v\} = \{1 + \epsilon u, u\}\{1 + \epsilon v, v\}.$$

Combining this with (6.35) gives

$$\{1 + \epsilon u, 1 + \epsilon v\} = 1 \text{ if } u, v, u + v, u - v \in F^*.$$

Since for any u, v we can write

$$\{1 + \epsilon u, 1 + \epsilon v\} = \prod_{i,j}\{1 + \epsilon u_i, 1 + \epsilon v_j\},$$

where $\sum_i u_i = u$, $\sum_j v_j = v$ and all $u_i, v_j, u_i + v_j, u_i - v_j \epsilon F^*$ we have the lemma. □

To prove (6.29) we shall show that

(i) $\{1 + \epsilon ac, a\}\{1 + \epsilon bc, b\} = \{1 + \epsilon c(a + b), a + b\}$,

(ii) $\{1 + \epsilon ab, a\}\{1 + \epsilon ab, b\} = \{1 + \epsilon ab, ab\}$,

(iii) $\{1+\epsilon a_1, b\}\{1+\epsilon a_2, b\} = \{1+\epsilon(a_1+a_2), b\}$,

(iv) $\{1+\epsilon a, \lambda\} = 1$ for $\lambda \in \mathbb{Q}^*$.

Here it is understood that $a, b, c \in F$ and elements of F are invertible when they need to be to make things well defined.

Assuming (i)–(iv), we have that the mapping

$$\ker\{K_2(F[\epsilon]) \to K_2(F)\} \to F \otimes_{\mathbb{Z}} F^*$$

given by

$$\{1+\epsilon a, b\} \to a \otimes b$$

is well-defined and injective, and (i)–(iv) generate exactly the relations that define the Kähler differentials; i.e., the kernel of the mapping

$$F \otimes_{\mathbb{Z}} F^* \longrightarrow \Omega^1_{F/\mathbb{Q}}$$

given by

$$a \otimes b \longrightarrow a\frac{db}{b}.$$

Thus the map induced by (6.33) is injective.

Since (ii) and (iii) are obvious it remains to prove (i) and (iv).

Let $A, A-1 \in F^*$ and $f \in F$. Then

$$\begin{aligned}
1 &= \left\{1 - \frac{A-1-\epsilon f}{A}, \frac{A-1-\epsilon f}{A}\right\} \\
&= \left\{\frac{1+\epsilon f}{A}, \frac{A-1-\epsilon f}{A}\right\} \\
&= \left\{\frac{1+\epsilon f}{A}, \frac{A-1}{A}\right\}\left\{\frac{1+\epsilon f}{A}, 1 - \frac{\epsilon f}{A-1}\right\} \\
&= \left\{1+\epsilon f, \frac{A-1}{A}\right\}\left\{\frac{1}{A}, \frac{A-1}{A}\right\}\left\{\frac{1}{A}, 1 - \frac{\epsilon f}{A-1}\right\}\left\{1+\epsilon f, 1-\frac{\epsilon f}{A-1}\right\}
\end{aligned}$$

The second term is

$$\left\{\frac{1}{A}, -\frac{1}{A}\right\}\left\{\frac{1}{A}, 1-A\right\} = 1$$

and the fourth term is also 1 by the second lemma above. We thus have

$$1 = \left\{1+\epsilon f, \frac{A-1}{A}\right\}\left\{A, 1+\frac{\epsilon f}{A-1}\right\},$$

where we have used $1+\frac{\epsilon f}{A-1} = (1-\frac{\epsilon f}{A-1})^{-1}$.

Now set $A = (a+b)/b$ so that $A-1 = a/b$, and set $f = ca$. Then $f/(A-1) = bc$ and

$$1 = \left\{1+\epsilon ac, \frac{a}{a+b}\right\}\left\{\frac{a+b}{b}, 1+\epsilon bc\right\}$$

which gives

$$\begin{aligned}
1 &= \{1+\epsilon ac, a\}\{1+\epsilon ac, a+b\}^{-1}\{a+b, 1+\epsilon bc\}\{b, 1+\epsilon bc\}^{-1} \\
&= \{1+\epsilon ac, a\}\{1+\epsilon ac, a+b\}^{-1}\{1+\epsilon bc, a+b\}^{-1}\{1+\epsilon bc, b\} \\
&= \{1+\epsilon ac, a\}\{1+\epsilon bc, b\}\{1+\epsilon c(a+b), a+b\}^{-1}
\end{aligned}$$

and this is (ii).

Proof of (iv): Using

$$\{1+\epsilon a, \lambda\} = \left\{1+\epsilon\frac{a}{m}, \lambda\right\}^m$$

and (6.31) we are reduced to showing

(6.36) $\{1+\epsilon a, p\}$ is torsion for p a prime.

Using that $\{\lambda, \mu\}$ is torsion for $\lambda, \mu \in \mathbb{Q}^*$ and computing modulo torsion, we have

$$\begin{aligned}\{1+\epsilon a, \mu\}^k &= \{\lambda, \mu\}^{-k}\{\lambda+\epsilon\lambda a, \mu\}^k \\ &= \{\lambda+\epsilon\lambda a, \mu(1-\lambda-\epsilon\lambda a)\}^k\end{aligned}$$

by the Steinberg relation

$$= \{\lambda+\epsilon\lambda a, \mu^k(1-\lambda)^k - k\mu^{k-1}(1-\lambda)^{k-1}\epsilon\lambda a\}.$$

Choosing $\lambda = 1/(1-k)$ gives

$$\begin{aligned}\{1+\epsilon a, \mu\}^k &= \{\lambda+\epsilon\lambda a, \mu^k(-k)^k\lambda^k - k\mu^{k-1}\,(-k)^{k-1}\lambda^k\epsilon a\} \\ &= \{\lambda+\epsilon\lambda a, \mu^k(-k)^k\lambda^k(1+\epsilon a)\} \\ &= \{\lambda, \mu^k(-k)^k\lambda^k\}\{\lambda, 1+\epsilon a\}\{1+\epsilon a, \mu^k(-k)^k\lambda^k\}\{1+\epsilon a, 1+\epsilon a\}.\end{aligned}$$

The first term is torsion, and the last one is 1 by the second lemma above. Thus, modulo torsion

$$\begin{aligned}\{1+\epsilon a, \mu\}^k &= \{1+\epsilon a, \mu^k\}\{1+\epsilon a, \lambda^{k-1}\}\{1+\epsilon k, (-k)^k\} \\ &= \{1+\epsilon a, \mu^k\}\left\{1+\epsilon a, \left(\frac{1}{k-1}\right)\right\}^{k-1}\{1+\epsilon a, k\}^k\end{aligned}$$

by our choice of λ

$$\Rightarrow \{1+\epsilon a, k\}^k = \{1+\epsilon a, (k-1)\}^{k-1}.$$

Now we may proceed inductively taking for k the smallest prime p for which $\{1+\epsilon a, p\}$ is not torsion. Using the prime power factorization of $p-1$ and bilinearity together with the induction assumption we arrive at a contradiction. This proves (6.36), and with it completes the proof of (6.29). □

Remark: An important invariant of $K_2(F)$ is given by the map

(6.37)
$$\begin{array}{ccc}\wedge^2 d\log : K_2(F) & \longrightarrow & \Omega^2_{F/\mathbb{Q}} \\ \Cup & & \Cup \\ \{a, b\} & \longrightarrow & \frac{da}{a}\wedge\frac{db}{b}.\end{array}$$

Replacing F by $F[\epsilon]$ and noting that

$$\ker\{\Omega^2_{F[\epsilon]/\mathbb{Q}} \to \Omega^2_{F/\mathbb{Q}}\} \cong d\epsilon\wedge\Omega^1_{F/\mathbb{Q}},$$

we have a commutative diagram

$$\begin{array}{ccc}TK_2(F) = \ker\{K_2(F[\epsilon]) \to K_2(F)\} & \xrightarrow{\wedge^2 d\log} & \ker\{\Omega^2_{F[\epsilon]/\mathbb{Q}} \to \Omega^2_{F/\mathbb{Q}}\} \\ \downarrow & & \downarrow \\ \Omega^1_{F/\mathbb{Q}} & \xrightarrow{\wedge d\epsilon} & d\epsilon\wedge\Omega^1_{F/\mathbb{Q}}\end{array}$$

This says that the tangent map factors through the $\wedge^2 d\log$ map (6.37). Taking, for example, $F = \mathbb{C}$ we have

$$\wedge^2 d\log\{a, b\} = 0 \Leftrightarrow a, b \text{ are algebraically dependent over } \mathbb{Q}.$$

This suggests that *information is lost in the tangent map for arithmetic reasons*. We will see this expressed more precisely when we discuss null curves in chapter 10 below.

We observe also that, unlike $TK_2(F)$, $T(F^* \otimes_{\mathbb{Z}} F^*)$ does not seem to have a particularly nice description. In particular for a, b nonzero

$$(1 + \epsilon a) \otimes (1 + \epsilon b) \in \ker\{F^*[\epsilon] \otimes_{\mathbb{Z}} F^*[\epsilon] \to F^* \otimes_{\mathbb{Z}} F^*\}$$

appears to be nonzero but in some sense to be of "order ϵ^2".

We have given the above argument to bring out the following geometric point. Let X be a smooth variety and

$$X_1 = X \times_{\mathbb{C}} \operatorname{Spec} \mathbb{C}[\epsilon]$$

the variety whose points are X and whose local rings are given by

$$\begin{aligned} \mathcal{O}_{X_1,x} &= \mathcal{O}_{X,x}[\epsilon] \\ &= \{f + \epsilon g : f, g \in \mathcal{O}_{X,x} \text{ and } \epsilon^2 = 0\}. \end{aligned}$$

Then, by definition, the tangent sheaf is

$$T\mathcal{K}_2(\mathcal{O}_X) = \ker\{\mathcal{K}_2(\mathcal{O}_{X_1}) \to \mathcal{K}_2(\mathcal{O}_X)\}$$

and by the above argument

$$T\mathcal{K}_2(\mathcal{O}_X) \cong \Omega^1_{X/\mathbb{Q}}.$$

Now suppose that X is defined over a field k that is finitely generated over $\mathbb{Q}$. The above argument works also for $X(k)$ and gives

$$T\mathcal{K}_2(\mathcal{O}_{X(k)}) \cong \Omega^1_{X(k)/\mathbb{Q}}.$$

On the other hand, in chapter 4 above we have given the interpretation (cf. (4.16d))

$$\Omega^1_{X(k)/\mathbb{Q},x} \cong \Omega^1_{\mathcal{X}(\mathbb{Q})/\mathbb{Q},(x,s_0)},$$

where $\mathcal{X} \to S$ is the k-spread of X. Combining these gives

$$T\mathcal{K}_2(\mathcal{O}_{X(k)}) \cong \Omega^1_{\mathcal{X}(\mathbb{Q})/\mathbb{Q}} \underset{k}{\otimes} \mathcal{O}_X \tag{6.38}$$

This will have the implication that

> *when considering algebraic cycles of higher codimension, the geometry of the spread necessarily enters.*

More specifically, "vertical" variations of cycle classes within X are measured by "horizontal" data in the spread of X and the cycle class. In chapter 10 below we shall explain how this geometric interpretation allows us to in some sense "integrate" Abel's differential equations.

Finally, we want to summarize some of the points above by discussing how K_2 and spreads *inevitably* arise from elementary geometric considerations, even if one is only interested in the geometry of *complex varieties*.

(a) The Weil reciprocity law, which is the exponentiated version of the residue theorem applied to the meromorphic differential

$$\log f \frac{dg}{g}$$

on the cut Riemannian surface, is expressed in terms of the tame symbol mappings

$$\mathbb{C}(X)^* \times \mathbb{C}(X)^* \xrightarrow{T_x} \mathbb{C}^*.$$

The kernel of T_x on

$$\left(\mathbb{C}(X)^*/\mathcal{O}_{X,x}^*\right) \times \left(\mathbb{C}(X)^*/\mathcal{O}_{X,x}^*\right)$$

is generated exactly by the relations that define $\mathbb{C}(X)^* \otimes_{\mathbb{Z}} \mathbb{C}(X)^*$ together with the Steinberg relation $T_x\left(f \otimes (1-f)\right) = 1$. Thus we have

$$0 \to \mathcal{K}_2(\mathcal{O}_X) \to \underline{\underline{K}}_2(\mathbb{C}(X)) \to \bigoplus_{x \in X} \underline{\underline{\mathbb{C}}}_x^* \to 1$$

and therefore once we consider the Weil reciprocity law K_2 is in the door.

(b) The above calculation shows that the Kähler differentials $\Omega^1_{X,x/\mathbb{Q}}$ arise when we "infinitesimalize" the relations that define $\mathcal{K}_2\left(\mathcal{O}_{X,x}[\epsilon]\right)$.

(c) As noted above, $\Omega^1_{\mathcal{O}_{X,x/\mathbb{Q}}}$ is interpreted as ordinary differential forms on spreads, once we keep track of the the field of definition.

(d) Finally, as is well known and will be discussed further in chapter 8 below, when localized the very definition of rational equivalence of 0-cycles on a surface X gives

$$\bigoplus_{\left\{\substack{\operatorname{codim} Y = 1 \\ Y \text{ irred}}\right.} \underline{\underline{\mathbb{C}}}(Y)^* \xrightarrow{\text{div}} \bigoplus_{x \in X} \underline{\underline{\mathbb{Z}}}_x \to 0$$

where div is the map $f \to \operatorname{div}(f)$ for $f \in \mathbb{C}(Y)^*$. The 2-dimensional version of (a) is that, locally in the Zariski topology, the kernel of div—the only way in which *cancellations* can occur—is in the image of the tame symbol

$$K_2(\underline{\underline{C}}(X)) \to \bigoplus_{Y} \underline{\underline{\mathbb{C}}}(Y)^*.$$

As in (a) the kernel of this map is $\mathcal{K}_2(\mathcal{O}_X)$, so that geometry inevitably leads to K_2. As in (b) and (c) this in turn inevitably leads to spreads.

Chapter Seven

The $\mathcal{E}xt$-definition of $TZ^2(X)$ for X an Algebraic Surface

In Chapter 5 we have given the geometric description

$$\underline{\underline{T}}Z^2(X) = \bigoplus_{x \in X} \underline{\underline{\mathrm{Hom}}}^o \left(\Omega^2_{X/\mathbb{Q},x}/\Omega^2_{\mathbb{C}/\mathbb{Q}}, \Omega^1_{\mathbb{C}/\mathbb{Q}}\right) \tag{7.1}$$

for the tangent space to the space of 0-cycles on a smooth algebraic surface.[1] This geometric description was in turn based on representing arcs in $Z^2(X)$ in local coordinates by Puiseaux series, taking the absolute differentials of these expressions, and then extracting what is invariant when one changes local uniformizing parameters. In invariant terms, one represents an arc in $Z^2(X)$ by a diagram

$$Z \subset X \times B, \qquad \dim B = 1,$$

and uses a pull-back, push-down construction to define the tangent map as a pairing

$$T_{s_0}B \otimes \left(\Omega^2_{X/\mathbb{Q},x}/\Omega^2_{\mathbb{C}/\mathbb{Q}}\right) \to \Omega^1_{\mathbb{C}/\mathbb{Q}}.$$

In either case we are giving arcs in the space of 0-cycles parametrically.

The other method to give such arcs is by their equations; i.e., by linear combinations of arcs in $\mathrm{Hilb}^2(X)$. This approach leads to what we feel is the proper definition of $\underline{\underline{T}}Z^2(X)$, one that although somewhat formal gives a better algebraic understanding of the tangent space and ties in more directly to duality theory. In section 7.1 we will give the formal definition of $\underline{\underline{T}}Z^2(X)$ and show that it agrees with the geometric description (7.1). In section 7.2 we will give the relation of $\underline{\underline{T}}Z^2(X)$ to $T\,\mathrm{Hilb}^2(X)$, and in section 7.3 we will give the direct comparison between the parameter (Puiseaux series) and equations ($\mathcal{E}xt^2$) approaches. Finally, in section 7.4 we will give some concluding remarks dealing with the desirable—but not yet accomplished—axiomatic approach to $\underline{\underline{T}}Z^2(X)$.

7.1 THE DEFINITION OF $\underline{\underline{T}}Z^2(X)$

Throughout this chapter, X will be a smooth algebraic surface, $x \in X$ a point, and Z a codimension-2 subscheme of X supported at x.

Definition: *The stalk at x of the tangent space to the space of 0-cycles is defined by*

$$\underline{\underline{T}}Z^2(X)_x = \lim_{\left\{\substack{Z \text{ codim } 2 \\ \text{subscheme} \\ \text{supported at } x}\right.} \mathcal{E}xt^2_{\mathcal{O}_X}\left(\mathcal{O}_Z, \Omega^1_{X/\mathbb{Q}}\right). \tag{7.2}$$

[1] Everything we shall do in this chapter extends to 0-cycles on an n-dimensional smooth variety. Since no essentially new ideas are involved in this extension, we shall restrict attention to the surface case.

Proposition: *We have the isomorphisms*

$$\underline{\underline{T}}Z^2(X)_x \cong H_x^2\left(\Omega^1_{X/\mathbb{Q}}\right) \cong \operatorname{Hom}^o\left(\Omega^2_{X/\mathbb{Q},x}/\Omega^2_{\mathbb{C}/\mathbb{Q}}, \Omega^1_{\mathbb{C}/\mathbb{Q}}\right). \tag{7.3}$$

It follows that the geometric description (7.1) and formal definition (7.2) agree *as vector spaces*. What remains to be done is to define the map

$$\begin{pmatrix}\text{linear combinations}\\ \text{of arcs in } \operatorname{Hilb}^2(X)\end{pmatrix} \to \underline{\underline{T}}Z^2(X),$$

where the stalks of $\underline{\underline{T}}Z^2(X)$ are given by (7.2) and show that when arcs in $\operatorname{Hilb}^2(X)$ are represented by Puiseaux series we obtain the same tangent maps under the identification (7.3). This will be done in the next two sections.

Before presenting the proof of the proposition, we remark that the limit in the right-hand side of (7.2) has the following meaning: If we have an inclusion

$$\mathcal{I}_{Z_1} \subset \mathcal{I}_{Z_2} \qquad \text{in } \mathcal{O}_{X,x}$$

inducing

$$\mathcal{O}_{Z_1} \to \mathcal{O}_{Z_2},$$

then there is an induced map

$$\mathcal{E}xt^2_{\mathcal{O}_X}\left(\mathcal{O}_{Z_2}, \Omega^1_{X/\mathbb{Q}}\right) \to \mathcal{E}xt^2_{\mathcal{O}_X}\left(\mathcal{O}_{Z_1}, \Omega^1_{X/\mathbb{Q}}\right)$$

and we use these maps to define the limit. If, as usual, $\mathfrak{m}_x$ denotes the maximal ideal in $\mathcal{O}_{X,x}$, then since Z is supported at x

$$\mathfrak{m}_x^k \subseteq \mathcal{I}_Z \quad \text{for } k \gg 0$$

and consequently

$$\lim_{\left\{\substack{Z \text{ codim 2}\\ \text{subscheme}\\ \text{supported at } x}\right.} \mathcal{E}xt^2_{\mathcal{O}_X}\left(\mathcal{O}_Z, \Omega^1_{X/\mathbb{Q}}\right) = \lim_k \mathcal{E}xt^2_{\mathcal{O}_X}\left(\mathcal{O}_X/\mathfrak{m}_x^k, \Omega^1_{X/\mathbb{Q}}\right). \tag{7.4}$$

Proof of proposition: For any scheme $Z \subset X$ as above, we have by Grothendieck duality (cf. section 8.2 for a summary of this)

$$\mathcal{E}xt^2_{\mathcal{O}_X}\left(\mathcal{O}_Z, \Omega^2_{X/\mathbb{C}}\right) \cong \operatorname{Hom}(\mathcal{O}_Z, \mathbb{C}).$$

From the exact sequence

$$0 \to \mathcal{O}_X \otimes \Omega^1_{\mathbb{C}/\mathbb{Q}} \to \Omega^1_{X/\mathbb{Q}} \to \Omega^1_{X/\mathbb{C}} \to 0$$

we then have

$$\begin{cases} \mathcal{E}xt^2_{\mathcal{O}_X}\left(\mathcal{O}_Z, \Omega^1_{X/\mathbb{C}}\right) \cong \operatorname{Hom}\left(\Omega^1_{X/\mathbb{C}} \otimes \mathcal{O}_Z, \mathbb{C}\right), \\ \mathcal{E}xt^2_{\mathcal{O}_X}\left(\mathcal{O}_Z, \mathcal{O}_X \otimes \Omega^1_{\mathbb{C}/\mathbb{Q}}\right) \cong \operatorname{Hom}\left(\Omega^2_{X/\mathbb{C}} \otimes \mathcal{O}_Z, \Omega^1_{\mathbb{C}/\mathbb{Q}}\right). \end{cases}$$

From Section 5.1 we recall the exact sequence

$$\begin{aligned} 0 \to \operatorname{Hom}(\Omega^2_{X/\mathbb{C}} \otimes \mathcal{O}_Z, \Omega^1_{\mathbb{C}/\mathbb{Q}}) &\to \operatorname{Hom}^o((\Omega^2_{X/\mathbb{Q}}/\Omega^2_{\mathbb{C}/\mathbb{Q}}) \otimes \mathcal{O}_Z, \Omega^1_{\mathbb{C}/\mathbb{Q}}) \\ &\to \operatorname{Hom}(\Omega^1_{X/\mathbb{C}} \otimes \mathcal{O}_Z, \mathbb{C}) \to 0. \end{aligned}$$

We thus have a diagram

$$\begin{array}{ccccccc}
0 \to \mathrm{Hom}\left(\Omega^2_{X/\mathbb{C}} \otimes \mathcal{O}_Z, \Omega^1_{\mathbb{C}/\mathbb{Q}}\right) & \to & \mathrm{Hom}^o\left(\left(\Omega^2_{X/\mathbb{Q}}/\Omega^2_{\mathbb{C}/\mathbb{Q}}\right) \otimes \mathcal{O}_Z, \Omega^1_{\mathbb{C}/\mathbb{Q}}\right) & \to & \mathrm{Hom}\left(\Omega^1_{X/\mathbb{C}} \otimes \mathcal{O}_Z, \mathbb{C}\right) \to 0 \\
\wr\| & & & & \wr\| \\
0 \to \mathcal{E}xt^2_{\mathcal{O}_X}\left(\mathcal{O}_Z, \mathcal{O}_X \otimes \Omega^1_{\mathbb{C}/\mathbb{Q}}\right) & \longrightarrow & \mathcal{E}xt^2_{\mathcal{O}_X}\left(\mathcal{O}_Z, \Omega^1_{X/\mathbb{Q}}\right) & \longrightarrow & \mathcal{E}xt^2_{\mathcal{O}_X}\left(\mathcal{O}_Z, \Omega^1_{X/\mathbb{Q}}\right) \longrightarrow 0.
\end{array}$$

If we can construct an upward vertical map in the middle compatible with the maps at the ends, then of necessity it is an isomorphism.

If

$$0 \to F_2 \to F_1 \to F_0 \to \mathcal{O}_Z \to 0$$

is a minimal free resolution (locally defined), and

$$F_2 \xrightarrow{\phi} \Omega^1_{X/\mathbb{Q}}$$

is an $\mathcal{O}_X$-linear map, then we have

$$\Omega^2_{X/\mathbb{Q}} \otimes F_2 \xrightarrow{\mathrm{id} \otimes \phi} \Omega^2_{X/\mathbb{Q}} \otimes \Omega^1_{X/\mathbb{Q}} \xrightarrow{\wedge} \Omega^3_{X/\mathbb{Q}} \to \Omega^2_{X/\mathbb{C}} \otimes \Omega^1_{\mathbb{C}/\mathbb{Q}}.$$

Thus ϕ gives a map

$$\begin{array}{c}
\Omega^2_{X/\mathbb{Q}} \to \mathrm{Hom}_{\mathcal{O}_X}\left(F_2, \Omega^2_{X/\mathbb{C}}\right) \otimes \Omega^1_{\mathbb{C}/\mathbb{Q}} \\
\downarrow \\
\mathcal{E}xt^2_{\mathcal{O}_X}\left(\mathcal{O}_Z, \Omega^2_{X/\mathbb{C}}\right) \otimes \Omega^1_{\mathbb{C}/\mathbb{Q}}.
\end{array}$$

There is a natural trace map

$$\mathcal{E}xt^2_{\mathcal{O}_X}\left(\mathcal{O}_Z, \Omega^2_{X/\mathbb{C}}\right) \to \mathrm{Hom}\left(\mathcal{O}_Z, \mathbb{C}\right) \to \mathbb{C}$$

where the right arrow is given by "evaluation on 1," and consequently ϕ gives a map

$$\Omega^2_{X/\mathbb{Q}} \longrightarrow \Omega^1_{\mathbb{C}/\mathbb{Q}}.$$

One checks easily that this annihilates $\Omega^2_{\mathbb{C}/\mathbb{Q}}$, does not depend on the choice of represntative ϕ of the original class in $\mathcal{E}xt^2_{\mathcal{O}_X}(\mathcal{O}_Z, \Omega^1_{X/\mathbb{Q}})$, and is compatible with the maps on the ends of the above diagram. □

We conclude this section with some considerations that lead to the following conclusion:

> *If one assumes (i) some "evident" continuity properties, and (ii) that the construction of the tangent map to cycles is "of an algebraic character" (cf. below for explanation), then the $\mathcal{E}xt$-definition (7.2) is uniquely determined.*

To begin with we pose the question:

If we have a family

$$z(t) = p_1(t) + \cdots + p_N(t)$$

where $p_1(t), \ldots, p_N(t)$ are distinct for $0 < |t| < \epsilon$, then one might ask:

How do we fill in over $t = 0$ the family of vector spaces

$$\bigoplus_{i=1}^{N} \Theta_{X/\mathbb{C}}\Big|_{p_i(t)} = \bigoplus_{i=1}^{N} T_{p_i(t)}X$$

so as to get a vector bundle over the parameter space in t?

If we note that for $0 < |t| < \epsilon$,

$$\mathcal{E}xt^2_{\mathcal{O}_X}\left(\mathcal{O}_{z(t)}, \Omega^1_{X/\mathbb{C}}\right) \cong \bigoplus_{i=1}^{N} \Theta_{X/\mathbb{C}}\Big|_{p_i(t)},$$

then we can answer the above question by filling in

$$\mathcal{E}xt^2_{\mathcal{O}_X}\left(\mathcal{O}_{z(0)}, \Omega^1_{X/\mathbb{C}}\right).$$

Since the first order derivative $z'(t)$ for $0 < |t| < \epsilon$ is certainly

$$\bigoplus_{i=1}^{N} p_i'(t) \in \bigoplus_{i=1}^{N} \Theta_{X/\mathbb{C}}\Big|_{p_i(t)},$$

it is a matter of taking the correct limit of the left-hand side as $t \to 0$ to get an element of

$$\mathcal{E}xt^2_{\mathcal{O}_X}\left(\mathcal{O}_{z(0)}, \Omega^1_{X/\mathbb{C}}\right).$$

This limit appears either in the Puiseaux construction or in the more algebraic construction of this section.

If, however, we take account of the fact that our variety X and family of cycles $z(t)$ are defined over a finitely generated field of definition k, then any geometrically meaningful construction should also spread out over k, and we thus should consider the spread $\mathcal{Z}(t)$, a cycle on the spread $\mathcal{X} \to S$ of X over k. Now the first order information lies in

$$\mathcal{E}xt^2_{\mathcal{O}_{\mathcal{X}}}\left(\mathcal{O}_{\mathcal{Z}(0)}, \Omega^1_{\mathcal{X}/\mathbb{C}}\right)$$

and this corresponds to

$$\mathcal{E}xt^2_{\mathcal{O}_X}\left(\mathcal{O}_{z(0)}, \Omega^1_{X/\mathbb{Q}}\right)$$

under the identification discussed previously.

This gives what we feel is a reasonably compelling explanation of why the formula for $\underline{\underline{T}}Z^2(X)$ takes the form that it does.

7.2 THE MAP $T\,\mathrm{Hilb}^2(X) \to TZ^2(X)$

In Chapter 5 we used the differentials of Puiseaux series, expressed in terms of a set of local uniformizing parameters, to construct for each point $x \in X$ a map

$$\left\{\begin{matrix}\text{arcs in } Z^2(X)\\ \text{starting at } x\end{matrix}\right\} \to \mathrm{Hom}^o\left(\Omega^2_{X/\mathbb{Q},x}/\Omega^2_{\mathbb{C}/\mathbb{Q}}, \Omega^1_{\mathbb{C}/\mathbb{Q}}\right)$$

$$\wr\|$$

$$\lim_{k\to\infty} \mathcal{E}xt^2_{\mathcal{O}_X}\left(\mathcal{O}_X/\mathfrak{m}_x^k, \Omega^1_{X/\mathbb{Q}}\right).$$

This was essentially a geometric construction, one which does not tell us how to go from a set of generators for the ideal defining the 0-cycle $z(t)$ to an element of $\mathcal{E}xt^2$. However, as will now be explained, based on a very nice construction of Angéniol and Lejeune-Jalabert [19] there is a direct way of doing this. Then in the next section we shall show that this construction agrees with the Puiseaux series method when the identification (7.3) is made.

We think of a family $z(t)$ of 0-dimensional subschemes as given by a subscheme

$$Z \subset X \times B,$$

where B is the parameter curve. We assume that $z(t)$ is supported in a neighborhood of $x \in X$ and that we have a family of free resolutions

$$0 \to F_2 \xrightarrow{f_2(t)} F_1 \xrightarrow{f_1(t)} F_0 \to \mathcal{O}_{z(t)} \to 0\,, \tag{7.5}$$

where if we wish we may think of each F_i locally as a direct sum of $\mathcal{O}_X$'s and $f_i(t)$ as a matrix with entries in $\mathcal{O}_{X\times B}$. By the composition

$$df_1(t) \circ df_2(t) : F_2 \to \Omega^2_{X\times B/\mathbb{Q}}$$

we mean "compose the matrices of absolute differentials using exterior product." Assuming that the support of $z(0)$ is x and letting $F_{i,x}$ be the stalk at x of F_i, we have the

Definition: *We define*

$$z'(0) \in \operatorname{Hom}_{\mathcal{O}_{X,x}}\left(F_{2,x}, \Omega^1_{X/\mathbb{Q},x}\right) \tag{7.6}$$

to be

$$z'(0) = \frac{1}{2}\,(df_1(t) \circ df_2(t)) \rfloor \frac{\partial}{\partial t}\Big|_{t=0}.$$

We will now show that this definition makes sense in

$$\mathcal{E}xt^2_{\mathcal{O}_X}\left(\mathcal{O}_Z, \Omega^1_{X/\mathbb{Q}}\right)\,.$$

First, if we write

$$f_1(t) = f_1 + t\dot{f}_1, \quad f_2 = f_2 + t\dot{f}_2 \qquad \bmod t^2$$

then

$$z'(0) = \frac{1}{2}\left(df_1 \circ \dot{f}_2 - \dot{f}_1 \circ df_2\right),$$

so that $z'(0)$ depends only on the first order deformation of $z(0)$.

However, $z'(0)$ does depend on the resolution (7.5). If we change resolutions according to a diagram

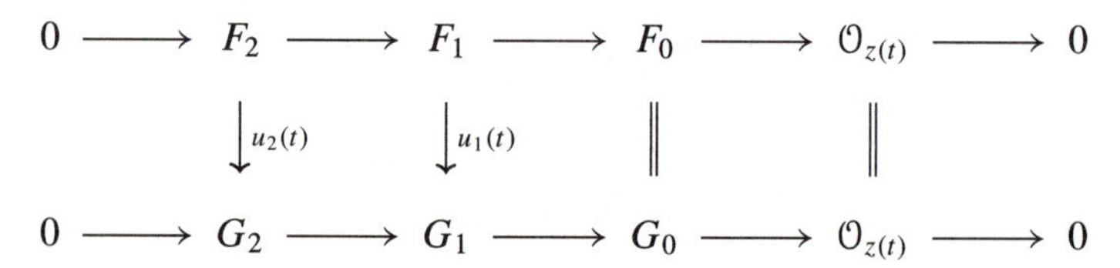

$$\begin{array}{ccccccccccc} 0 & \longrightarrow & F_2 & \longrightarrow & F_1 & \longrightarrow & F_0 & \longrightarrow & \mathcal{O}_{z(t)} & \longrightarrow & 0 \\ & & \big\downarrow{\scriptstyle u_2(t)} & & \big\downarrow{\scriptstyle u_1(t)} & & \| & & \| & & \\ 0 & \longrightarrow & G_2 & \longrightarrow & G_1 & \longrightarrow & G_0 & \longrightarrow & \mathcal{O}_{z(t)} & \longrightarrow & 0 \end{array}$$

where $F_0 = G_0 = \mathcal{O}_X$, then denoting by $g_i(t)$ the maps on the bottom and using that the $u_i(t)$ are invertible we have

$$\begin{aligned} dg_1(t) \circ dg_2(t) &= d(f_1(t) \circ u_1(t)^{-1}) \circ \big(du_1(t) \circ f_2(t) \circ u_2(t)^{-1}\big) \\ &= \big(df_1(t) \circ u_1(t)^{-1} - f_1(t) \circ u_1(t)^{-1} \circ du_1(t) \circ u_1(t)^{-1}\big) \\ &\quad \circ \big(du_1(t) \circ f_2(t) \circ u_2(t)^{-1} + u_1(t) \circ df_2(t) \circ u_2(t)^{-1} \\ &\quad - u_1(t) \circ f_2(t) \circ u_2(t)^{-1} \circ du_2(t) \circ u_2(t)^{-1}\big). \end{aligned}$$

By passing to $\mathcal{E}xt^2$, these ambiguities will drop out. Indeed, it is a general fact that

$$\mathcal{I}_Z \text{ annihilates } \mathcal{E}xt^2_{\mathcal{O}_X}(\mathcal{O}_Z, \mathcal{F})$$

for any coherent sheaf of $\mathcal{O}_X$-modules $\mathcal{F}$, and applying this for $z(t)$, the

$$f_1(t) \circ u_1(t)^{-1} \circ du_1(t) \circ u_1(t)^{-1}$$

term drops out. The

$$du_1(t) \circ f_2(t) \circ u_2(t)^{-1}$$

term drops out by the definition of $\mathcal{E}xt^2$. So in $\mathcal{E}xt^2_{\mathcal{O}_X}\left(\mathcal{O}_Z, \Omega^1_{X \times B/\mathbb{Q}}\right)$ we are left with

$$\begin{aligned} dg_1(t) \circ dg_2(t) = &\ df_1(t) \circ df_2(t) \circ u_2(t)^{-1} \\ &- df_1(t) \circ f_2(t) \circ u_2(t)^{-1} \circ du_2(t) \circ u_2(t)^{-1}. \end{aligned}$$

The last term on the right drops out because, since from

$$f_1(t) \circ f_2(t) = 0$$

we have

$$f_1(t) \circ df_2(t) = -df_1(t) \circ f_2(t).$$

Consequently the problem term becomes

$$f_1(t) \circ df_2(t) \circ u_2(t)^{-1} \circ du_2(t) \circ u_2(t)^{-1},$$

which drops out because $\mathcal{I}_{z(t)}$ annihilates the $\mathcal{E}xt$ group. Thus in $\mathcal{E}xt^2$

$$dg_1(t) \circ dg_2(t) = df_1(t) \circ df_2(t) \circ u_2(t)^{-1}$$

and we understand the definition (7.6) to be

$$z'(0) = \frac{1}{2}\,(df_1(t) \circ df_2(t)) \rfloor \frac{\partial}{\partial t}\bigg|_{t=0} \in \mathcal{E}xt^2_{\mathcal{O}_X}\left(\mathcal{O}_Z, \Omega^1_{X/\mathbb{Q}}\right), \tag{7.7}$$

where $Z = z(0)$.

We may view this construction as giving a map

$$T\,\mathrm{Hilb}^2(X) \to \underline{\underline{T}}Z^2(X).$$

Example: $z(t) = (at, bt), \quad a, b \in \mathbb{C}$.

The resolution is just

$$0 \to \mathcal{O}_X \xrightarrow{\begin{pmatrix} y-bt \\ -(x-at) \end{pmatrix}} \mathcal{O}_X^2 \xrightarrow{(x-at, y-bt)} \mathcal{O}_X \to \mathcal{O}_{z(t)} \to 0$$

and

$$\frac{1}{2} df_1(t) \circ df_2(t) = \frac{1}{2} (dx - adt - tda \quad dy - bdt - tdb) \circ \begin{pmatrix} dy - bdt - tdb \\ -dx + adt + tda \end{pmatrix}$$
$$= dx - dy + (ady - bdx) \wedge dt \qquad \text{mod } t.$$

So

$$z'(0) = ady - bdx.$$

Here, upon choosing this resolution we have an isomorphism

$$\mathcal{E}xt^2_{\mathcal{O}_X}(\mathcal{O}_Z, \Omega^1_{X/\mathbb{Q}}) \cong \Theta_{X/\mathbb{Q}} \otimes_{\mathcal{O}_X} \mathcal{O}_Z$$

and $z'(0)$ is the element corresponding to a lifting of the tangent vector (a, b).

Example: $I_{z(t)} = (x^2, y + tx)$. Note that as a scheme $z(t)$ is varying, but as a cycle it is $2 \cdot (0, 0)$ for all t. We thus expect to find $z'(0) = 0$.

The resolution is

$$0 \to \mathcal{O}_X \xrightarrow{\begin{pmatrix} y+tx \\ -x^2 \end{pmatrix}} \mathcal{O}_X^2 \xrightarrow{(x^2 \ y+tx)} \mathcal{O}_X \to \mathcal{O}_{z(t)} \to 0$$

and

$$\frac{1}{2} df_1(t) \circ df_2(t) = 2xdx \wedge (dy + tdx + xdt)$$
$$= 2xdx \wedge dy + 2x^2 dx \wedge dt.$$

So,

$$z'(0) = x^2 dx.$$

However, using this resolution,

$$\mathcal{E}xt^2_{\mathcal{O}_X}\left(\mathcal{O}_Z, \Omega^1_{X/\mathbb{Q}}\right) = \left(\mathcal{O}_X/(x^2, y)\right) \otimes \Omega^1_{X/\mathbb{Q}}$$

and consequently

$$x^2 dx = 0,$$

which gives

$$z'(0) = 0$$

as expected.

We are now ready for a more subtle example that illustrates one of the most novel elements of the construction, namely the use of differentials over $\mathbb{Q}$.

Example: $I_{z_\alpha(t)} = (x^2 - \alpha y^2, xy - t), \quad \alpha \in \mathbb{C}^*$.

This example was discussed at length in section 6.1, and we want to recover its properties using $z'(0)$. The resolution is

$$0 \to \mathcal{O}_X \xrightarrow{\binom{x^2-\alpha y^2}{-xy+t}} \mathcal{O}_X \xrightarrow{(xy-t, x^2-\alpha y^2)} \mathcal{O}_X \to \mathcal{O}_{z_\alpha(t)} \to 0$$

and a computation as above gives

$$z'_\alpha(0) = 2xdx - 2\alpha ydy - y^2 d\alpha.$$

In order to compare $z'_\alpha(0)$ for different α's, we go to the group

$$\lim_{\left\{\substack{Z \text{ supported} \\ \text{at } (0,0)}\right.} \mathcal{E}xt^2_{\mathcal{O}_X}\left(\mathcal{O}_Z, \Omega^1_{X/\mathbb{Q}}\right).$$

We note that if $\mathfrak{m}$ is the maximal ideal at $(0, 0)$, then

$$I_{z_\alpha(0)} \supseteq \mathfrak{m}^3 \quad \text{for all } \alpha.$$

We can thus lift $z'_\alpha(0)$ to

$$\mathcal{E}xt^2_{\mathcal{O}_X}\left(\mathcal{O}_X/\mathfrak{m}^3, \Omega^1_{X/\mathbb{Q}}\right)$$

and then compare them for different α's. We have the commutative diagram

$$\begin{array}{ccccccccc}
0 \to \mathcal{O}_X & \xrightarrow{\binom{xy}{\alpha y^2 - x^2}} & \mathcal{O}_X^2 & \xrightarrow{(x^2-\alpha y^2\ xy)} & \mathcal{O}_X & \to & \mathcal{O}_{z_\alpha(0)} & \to & 0 \\
\uparrow{\scriptstyle (1\ 1/\alpha)} & & \uparrow{\scriptstyle \begin{pmatrix} x & 0 & 0 & -y/\alpha \\ \alpha y & x & y & x/\alpha \end{pmatrix}} & & \| & & \uparrow & & \\
0 \to \mathcal{O}_X^3 & \xrightarrow{\begin{pmatrix} y & 0 & 0 \\ -x & y & 0 \\ 0 & -x & y \\ 0 & 0 & -x \end{pmatrix}} & \mathcal{O}_X^4 & \xrightarrow{(x^3\ x^2y\ xy^2\ y^3)} & \mathcal{O}_X & \to & \mathcal{O}_X/\mathfrak{m}^3 & \to & 0.
\end{array}$$

So $z'_\alpha(0)$ pulls back to

$$\left(2xdx - 2\alpha ydy - y^2 d\alpha \quad 0 \quad \frac{2x}{\alpha}dx - 2ydy - y^2\frac{d\alpha}{\alpha}\right).$$

In

$$\mathcal{E}xt^2_{\mathcal{O}_X}\left(\mathcal{O}_X/\mathfrak{m}^3, \Omega^1_{X/\mathbb{Q}}\right)$$

we have the relations

$$(y\,0\,0), (-x\ y\ 0), (0\ -x\ y), (0\,0\ -x).$$

Then $z'_\alpha(0)$ is equivalent to

$$\left(2xdx \quad 0 \quad -2ydy - y^2\frac{d\alpha}{\alpha}\right).$$

Thus

$$z'_\alpha(0) - z'_1(0) = \left(0 \quad 0 \quad -y^2\tfrac{d\alpha}{\alpha}\right).$$

Since

$$(0, 0, y^2) \neq 0 \text{ in } \mathcal{E}xt^2_{\mathcal{O}_X}\left(\mathcal{O}_X/\mathfrak{m}^3, \mathcal{O}_X\right)$$

we see that

$$\begin{aligned} z'_\alpha(0) - z'_1(0) = 0 &\Leftrightarrow \frac{d\alpha}{\alpha} = 0 \\ &\Leftrightarrow \alpha \text{ is algebraic.} \end{aligned}$$

This is an exact fit for the discussion in section 6.1.

Remark: The construction given above makes sense in other dimensions and codimensions. For codimension $n = \dim X$, it goes through exactly, but for codimension $< n$ and Z not locally Cohen-Macaulay, the resolution may have the wrong number of steps and the construction requires modification.

7.3 RELATION OF THE PUISEAUX AND ALGEBRAIC APPROACHES

In geometry a submanifold—or more generally a subvariety—may be given "parametrically" either by a map $f : M \to N$ or by the equations that define $f(M) \subset N$. The Puiseaux series and $\mathcal{E}xt^2$ approaches to $\underline{\underline{T}} Z^2(X)$ reflect these two perspectives. Each has its virtues and drawbacks. An obvious next issue for this study is to show that they coincide. Before doing this we mention that on the one hand the Puiseaux approach has the following desirable features:

- It is clearly additive.
- It depends only on $z(t)$ as a cycle.
- It depends on $z(t)$ only up to first order in t.
- It has clear geometric meaning.

It has the undesirable feature:

- Two families of effective cycles might represent the same element of $T\,\mathrm{Hilb}^2(X)$ but not give the same tangent under the Puiseaux approach.

On the other hand the algebraic approach has the following desirable features:

- It clearly depends only on the element of $T\,\mathrm{Hilb}^2(X)$ determined by $z(t)$.
- It is easy to compute in examples.

It has the undesirable features:

- Additivity does not make sense for arbitrary schemes.
- It is not clear that $z'(0)$ depends only on the underlying cycle structure of $z(t)$.

In order to have the best of both approaches as well as, of course, for the development of the theory, we need to know that the two maps coincide; i.e., that the diagram

$$\begin{array}{ccc} \{\text{arcs in } Z^2(X)\} & \xrightarrow{\binom{\text{Puiseaux}}{\text{approach}}} & \bigoplus_{x\in X} \operatorname{Hom}^o\left(\Omega^2_{X/\mathbb{Q},x}/\Omega^2_{\mathbb{C}/\mathbb{Q}}, \Omega^1_{\mathbb{C}/\mathbb{Q}}\right) \\ \uparrow & & \wr\| \\ \{\text{arcs in } \operatorname{Hilb}^2(X)\} & \xrightarrow{\binom{\text{algebraic}}{\text{approach (7.7)}}} & \lim_{\{Z \text{ codim 2 subscheme}} \mathcal{E}xt^2_{\mathcal{O}_X}\left(\mathcal{O}_Z, \Omega^1_{X/\mathbb{Q}}\right) \end{array} \tag{7.8}$$

commutes where the right-hand vertical isomorphism is (7.2).

Proof that (7.8) commutes: The strategy is

Step 1: Verify equality in the case of a single moving point.

Step 2: Verify equality when $z(t)$ consists of distinct points.

Step 3: Pass from distinct points to arbitrary 0-dimensional schemes by taking limits.

We shall give the calculations in local coordinates x, y, which may be thought of as representatives of local uniformizing parameters in the completion of the local ring of the algebraic surface X at a closed point.

Proof of Step 1: Let

$$z(t) = (x_0 + at, y_0 + bt).$$

The Puiseaux expansion method has

$$dx \wedge dy = (dx_0 + adt + tda) \wedge (dy_0 + bdt + tdb),$$

which, at $t = 0$, has dt coefficient $bdx_0 - ady_0$, so that

$$dx \wedge dy \mapsto bdx_0 - ady_0.$$

Similarly

$$\begin{aligned} (x-x_0)^i(y-y_0)^j dx \wedge dy &\mapsto 0 && \text{if } i > 0 \text{ or } j > 0, \\ dx \wedge d\alpha &\mapsto -ad\alpha, && \text{all } \alpha \in \mathbb{C}, \\ dy \wedge d\alpha &\mapsto -bd\alpha, && \text{all } \alpha \in \mathbb{C}, \\ (x-x_0)^i(y-y_0)^j dx \wedge d\alpha &\mapsto 0 && \text{if } i > 0 \text{ or } j > 0, \\ (x-x_0)^i(y-y_0)^j dy \wedge d\alpha &\mapsto 0 && \text{if } i > 0 \text{ or } j > 0. \end{aligned}$$

The algebraic construction gives

$$0 \to \mathcal{O}_X \xrightarrow{\begin{pmatrix} y-y_0-bt \\ -(x-x_0-at) \end{pmatrix}} \mathcal{O}_X^2 \xrightarrow{(x-x_0-at \;\; y-y_0-bt)} \mathcal{O}_X \to \mathcal{O}_{z(t)} \to 0$$

and

$$\frac{1}{2} df_1 \circ df_2 = (dx - dx_0 - adt - tda) \wedge (dy - dy_0 - bdt - tdb);$$

consequently

$$z'(0) = -b(dx - dx_0) + a(dy - dy_0).$$

The trace map

$$\mathcal{E}xt^2_{\mathcal{O}_X}\left(\mathcal{O}_Z, \Omega^2_{X/\mathbb{C}}\right) \xrightarrow{\mathrm{Tr}} \mathbb{C}$$

for this resolution is

$$dx \wedge dy \mapsto 1.$$

To identify $z'(0)$ with an element of

$$\mathrm{Hom}^o\left(\left(\Omega^2_{X/\mathbb{Q}}/\Omega^2_{\mathbb{C}/\mathbb{Q}}\right) \otimes \mathcal{O}_Z, \Omega^1_{\mathbb{C}/\mathbb{Q}}\right)$$

the prescription is that we send

$$\omega \mapsto (z'(0) \wedge \omega) \rfloor \partial/\partial x \wedge \partial/\partial y \in \Omega^1_{\mathbb{C}/\mathbb{Q}}.$$

It follows that

$$\begin{aligned} dx \wedge dy \mapsto{}& ((-b(dx - dx_0) + a(dy - dy_0)) \wedge dx \wedge dy) \rfloor \partial/\partial x \wedge \partial/\partial y \\ &= b dx_0 - a dy_0, \\ dx \wedge d\alpha \mapsto{}& ((-b(dx - dx_0) + a(dy - dy_0)) \wedge dx \wedge d\alpha) \rfloor \partial/\partial x \wedge \partial/\partial y \\ &= -a d\alpha, \\ dy \wedge d\alpha \mapsto{}& ((-b(dx - dx_0) + a(dy - dy_0)) \wedge dy \wedge d\alpha) \rfloor \partial/\partial x \wedge \partial/\partial y \\ &= -b d\alpha. \end{aligned}$$

All forms with an $x^i y^j$, $i > 0$ or $j > 0$, go to zero. So the two constructions agree in this case.

Proof of Step 2:

Lemma: *If we use a local free resolution of the form*

$$0 \to F_2 \xrightarrow{f_2(t)} F_1 \xrightarrow{f_1(t)} F_0 \to \mathcal{O}_{z(t)} \to 0,$$

even if it is a nonminimal resolution, it gives the correct answer for the algebraic construction.

Proof of Lemma: If

$$0 \to F_2 \xrightarrow{f_2} F_1 \xrightarrow{f_1} F_0 \to \mathcal{O}_Z \to 0$$

is a minimal local free resolution and

$$0 \to F_2 \oplus A \xrightarrow{\begin{pmatrix} f_2 & 0 \\ 0 & u_2 \\ 0 & 0 \end{pmatrix}} F_1 \oplus A \oplus B \xrightarrow{\begin{pmatrix} f_1 & 0 & 0 \\ 0 & 0 & u_1 \end{pmatrix}} F_0 \oplus B \to \mathcal{O}_Z \to 0$$

where u_1, u_2 have constant entries on X (but possibly varying in t), then

$$\begin{aligned} d\begin{pmatrix} f_1 & 0 & 0 \\ 0 & 0 & u_1 \end{pmatrix} \circ \begin{pmatrix} \dot{f}_2 & 0 \\ 0 & \dot{u}_2 \\ 0 & 0 \end{pmatrix} - \begin{pmatrix} \dot{f}_1 & 0 & 0 \\ 0 & 0 & \dot{u}_1 \end{pmatrix} \circ d\begin{pmatrix} f_2 & 0 \\ 0 & u_2 \\ 0 & 0 \end{pmatrix} \\ = \begin{pmatrix} df_1 \circ \dot{f}_2 - \dot{f}_1 \circ df_2 & 0 \\ 0 & 0 \end{pmatrix}. \end{aligned}$$

This proves the lemma. □

If

$$z(t) = p_1(t) + \cdots + p_N(t)$$

with $p_1, \ldots, p_N$ distinct, then

$$\mathcal{E}xt^2_{\mathcal{O}_X}\left(\mathcal{O}_{z(t)}, \Omega^1_{X/\mathbb{Q}}\right) \cong \bigoplus_{i=1}^{N} \mathcal{E}xt^2_{\mathcal{O}_X}\left(\mathcal{O}_{p_i(t)}, \Omega^1_{X/\mathbb{Q}}\right).$$

Since the sheaves on the right are supported at distinct points, there are canonical maps going both ways in this identification. Whatever local minimal free resolution is used for $\mathcal{O}_{z(t)}$, on a suitable neighborhood of $p_i(t)$ it becomes a (nonminimal) local free resolution of $\mathcal{O}_{p_i(t)}$. By the lemma, the algebraic construction still gives the correct answer.

Additivity is now automatic, since pulling back from each

$$\mathcal{E}xt^2_{\mathcal{O}_X}\left(\mathcal{O}_{p_i(t)}, \Omega^1_{X/\mathbb{Q}}\right)$$

to

$$\mathcal{E}xt^2_{\mathcal{O}_X}\left(\mathcal{O}_Z, \Omega^1_{X/\mathbb{Q}}\right)$$

and summing then commutes with summing in

$$\bigoplus_{i=1}^{N} \mathcal{E}xt^2_{\mathcal{O}_X}\left(\mathcal{O}_{p_i(t)}, \Omega^1_{X/\mathbb{Q}}\right)$$

followed by the canonical map to $\mathcal{E}xt^2_{\mathcal{O}_X}\left(\mathcal{O}_Z, \Omega^1_{X/\mathbb{Q}}\right)$. This completes the proof of this step.

Proof of Step 3: We can make any family of 0-dimensional subschemes $z(t)$ part of a 2-parameter family $z(t, u)$, where

$$z(t) = z(t, 0) \quad \text{for all } t$$

and for some $\epsilon > 0$,

$$z(t, u) \text{ consists of distinct points for } 0 < |u| < \epsilon.$$

By flatness considerations, the dimensions of F_2, F_1 and $\mathcal{E}xt^2_{\mathcal{O}_X}\left(\mathcal{O}_{z(t,u)}, \mathcal{O}_X\right)$ are constant at t, u. If

$$z_u(t) = z(t, u)$$

then

$$\lim_{u \to 0} z'_u(0) = z'(0).$$

Also, for a fixed $\omega \in \Omega^2_{X/\mathbb{Q}}$, for the Puiseaux construction it is clear that

$$\lim_{u \to 0} \langle \omega, z_u(t) \rangle = \langle \omega, z(t) \rangle.$$

We thus pass from equality of the two constructions for a family with distinct points to an arbitrary family.

To complete the argument we need to provide some clarification about taking limits in $\Omega^1_{X/\mathbb{Q}}$. In the above we want to have

$$\lim_{u \to 0} d\alpha(u) = d\alpha(0)$$

while avoiding the problem that in general for a sequence $\alpha_i \in \mathbb{C}$

$$\lim_{i \to 0} d\alpha_i \neq d(\lim_{i \to 0} \alpha_i).$$

The correct formalism here is to use spreads, as in section 4.2. We let $k \subseteq \mathbb{C}$ be a finitely generated field over $\mathbb{Q}$ over which x and $z(t)$ are defined, S a smooth variety defined over $\mathbb{Q}$ and with $\mathbb{Q}(S) \cong k$, and $s_0 \in S$ the point corresponding to the given embedding of k in $\mathbb{C}$. Let

$$\begin{array}{c} \mathcal{X} \\ \downarrow \\ S \end{array}$$

be the spread of X over S. We may now use $\Omega^1_{\mathcal{X}/\mathbb{C}}$ rather than $\Omega^1_{X/\mathbb{Q}}$ and take limits there in the usual sense.

Remark: We have remarked that there is no natural way to assign a scheme structure to a 0-cycle. If, however, we have a family $z(t)$ of 0-dimensional subschemes giving rise to a family of cycles, then the assertion that the tangent map to cycles factors

$$T\,\mathrm{Hilb}^2(X) \to TZ^2(X)$$

asserts that $\mathcal{I}_{z(0)}$ together with

$$\dot{\mathcal{I}}_{z(0)} \stackrel{\text{defn}}{=} \left.\frac{d\,\mathcal{I}_{z(t)}}{dt}\right|_{t=0} : \mathcal{I}_{z(0)} \to \mathcal{O}_{X,z(0)}$$

determine the tangent vector $z'(0)$ to the space of cycles arising from $z(t)$ considered as an arc in $Z^2(X)$. In particular, $\mathcal{I}_{z(0)}$ and $\dot{\mathcal{I}}_{z(0)}$ determine how $z'(0)$ acts on the 1-forms $\Omega^1_{X/\mathbb{C},x}$ and 2-forms $\Omega^2_{X/\mathbb{C},x}$. At first this seemed surprising to us since it means that somehow some absolute differentials are determined by the geometric quantities $\mathcal{I}_{z(0)}$ and $\dot{\mathcal{I}}_{z(0)}$.

For example, the family of cycles given by Puiseaux series

$$z(t) = (a_1 t^{1/2} + a_2 t, b_1 t^{1/2} + b_2 t) + (-a_1 t^{1/2} + a_2 t, -b_1 t^{1/2} + b_2 t)$$

lifts to the family of schemes

$$z(t) = \mathrm{Var}(b_1 x - a_1 y + (a_1 b_2 - a_2 b_1)t,\ x^2 - 2a_2 xt + a_1^2 t) \bmod t^{3/2}.$$

Note that

$$\mathcal{I}_{z(0)} = (b_1 x - a_1 y, x^2)$$

determines

$$a_1/b_1$$

and $z'(0)$ is the map

$$\begin{aligned} b_1x - a_2y &\mapsto a_1b_2 - a_2b_1 \\ x^2 &\mapsto -2a_2x + a_1^2. \end{aligned}$$

From this information, we can recover $(a_1^2, a_1b_1, b_1^2, a_2, b_2)$ and hence the original Puiseaux series—so, a fortiori, the information obtained by the action on differential forms.

Of course, for a sum of Puiseaux series of this type, one gets a much more complicated relationship between $\mathcal{I}_{z(0)}, \dot{\mathcal{I}}_{z(0)}$ and the coefficients of the various Puiseaux series. What the factorization result asserts is that this information is always enough to allow us to recover the action of differential forms on the underlying sum of Puiseaux series. For example, for

$$z(t) = \sum_{\nu=1}^{N} \left((a_1^\nu t^{1/2} + a_2^\nu t, b_1^\nu t^{1/2} + b_2^\nu t) + (-a_1^\nu t^{1/2} + a_2^\nu t, -b_1^\nu t^{1/2} + b_2^\nu t\right),$$

the action of $dx \wedge dy$ is

$$z'(0)(dx \wedge dy) = \sum_{\nu=1}^{N} (a_1^\nu db_1^\nu - b_1^\nu da_1^\nu) \in \Omega^1_{\mathbb{C}/\mathbb{Q}}.$$

The theorem asserts that one can recover this from knowing $\mathcal{I}_{z(0)}, \dot{\mathcal{I}}_{z(0)}$, but because the formula for the ideal $\mathcal{I}_{z(t)}$ is difficult to write down, the exact formula for how to recover this invariant will depend on N. The good aspect of the

$$\frac{1}{2}(df_1 \cdot \dot{f}_2 - \dot{f}_1 \circ df_2)$$

formula is that it gives the correct answer without needing to know formulas for the relationship between the representation as an ideal and the representation as a Puiseaux series.

7.4 FURTHER REMARKS

Combining the virtues of the algebraic and Puiseaux constructions, we now have for a surface X a well-defined group homomorphism

$$\left\{\begin{array}{c}\text{Free group on}\\ \text{1-parameter families}\\ \text{of 0-dimensional}\\ \text{subschemes of } X\end{array}\right\} \xrightarrow{\tau} \lim_{\left\{\substack{Z \text{ codim } 2\\ \text{subschemes}}\right.} \mathcal{E}xt^2_{\mathcal{O}_X}\left(\mathcal{O}_Z, \Omega^1_{X/\mathbb{Q}}\right) \cong \bigoplus_{x\in X} H^2_x(\Omega^1_{X/\mathbb{Q}})$$

having the properties:

(i) τ depends only on the underlying cycle of $\sum_i z_i(t)$.

(ii) $\tau(z(t))$ depends only on the tangent vector to $z(t)$ in the Hilbert scheme; i.e., on $dI_{z(t)}/dt\ |_{t=0} \mod I_{z(0)}$.

(iii) τ kills torsion; i.e., if $\tau(nz(t)) = 0$ then $\tau(z(t)) = 0$.

(iv) τ is intrinsic and functorial, compatible with push-forward and pullback for maps of surfaces.

(v) τ behaves well under limits; i.e., if $z(t, u)$ is a 2-parameter family of 0-cycles and $z_u(t) = z(t, u)$, then

$$\lim_{u \to 0} \tau(z_u(t)) = \tau(z_0(t)).$$

There is also the property that τ is functorial under maps to $Y \to X$ and $X \to Y$, where Y is a curve and we use the tangent space to 0-cycles on curves described in chapter 2.

We would expect that τ is universal in the sense that any group homomorphism satisfying properties (i)–(v) factors through τ. In particular, we would expect $\ker(\tau)$ to be the smallest subgroup of the free group on 1-parameter families of 0-dimensional subschemes of X that satisfies the consequences of (i)–(v). We have been unable to prove this latter statement and consider it an important open problem — some evidence for how things might go is the result and argument from section 6.1 about the families $z_{\alpha\beta}(t)$ with

$$I_{z_{\alpha,\beta}(t)} = (x^2 - \alpha y^2, xy - \beta t),$$

where the element

$$\beta \frac{d\alpha}{\alpha} \in \Omega^1_{\mathbb{C}/\mathbb{Q}}$$

completely controls the situation, and this fact can be reduced to basic geometric equivalences.

In concluding this chapter, we would like to make a direct connection between our original "computational" approach to tangents to arcs in $Z^2(X)$ by taking differentials of Puiseaux series, and the intrinsic $\mathcal{E}xt$-approach just given. The first key observation is that, given a point $x \in X$ and local uniformizing parameters $\xi, \eta \in \mathbb{C}(X)$ centered at x, we have by section 8.2 there is a well-known identification

$$\lim_{k \to \infty} \mathcal{E}xt^2_{\mathcal{O}_X}\left(\mathcal{O}_X/\mathfrak{m}_x^k, \Omega^1_{X/\mathbb{Q}}\right) \cong \left\{ \begin{array}{c} 4^{\text{th}} \text{ quadrant Laurent} \\ \text{tails } \sum\limits_{i,j>0} \alpha_{ij}/\xi^i\eta^j \\ \text{where } \alpha_{ij} \in \Omega^1_{X/\mathbb{Q}}\big|_x \end{array} \right\}. \tag{7.9}$$

Here the notation $\Omega^1_{X/\mathbb{Q}}\big|_x$ means $\Omega^1_{X/\mathbb{Q},x}/\mathfrak{m}_x\Omega^1_{X/\mathbb{Q},x}$. Given an arc $z(t)$ in $Z^1_{\{x\}}(X)$— i.e., with $\lim_{t\to\infty} |z(t)| = x$—we want to identify its tangent $z'(0)$ as given by a fourth quadrant Laurent tail with coefficients in $\Omega^1_{X/\mathbb{Q}}\big|_x$, as above. The key is to use the canonical element

$$\omega = \frac{d\xi \wedge d\eta}{\xi\eta} \in \mathcal{E}xt^2_{\mathcal{O}_X}\left(\mathcal{O}_X/\mathfrak{m}_x, \Omega^2_{X/\mathbb{C}}\right) \tag{7.10}$$

(the sense in which this element is canonical will also be discussed below). Using the local uniformizing parameters ξ, η we write (as in chapter 3)

$$z(t) = x_1(t) + \cdots + x_m(t)$$

as a sum of Puiseaux series

$$x_i(t) = (\xi_i(t), \eta_i(t)) .$$

Claim: *If we write*

$$\sum_i \frac{d(\xi - \xi_i(t)) \wedge d(\eta - \eta_i(t))}{(\xi - \xi_i(t))(\eta - \eta_i(t))} \equiv \varphi \wedge dt + \gamma \mod t, \tag{7.11}$$

where γ does not involve dt, then φ is a fourth quadrant Laurent series with coefficients in $\Omega^1_{X/\mathbb{Q}}\big|_x$ whose coefficients are exactly the terms that arise as differentials of Puiseaux series as in chapter 3.

More precisely, if we consider

$$z'(0) \in \mathrm{Hom}^o\left(\Omega^2_{X/\mathbb{Q},x}/\Omega^2_{\mathbb{C}/\mathbb{Q}}, \Omega^1_{\mathbb{C}/\mathbb{Q}}\right)$$

as in the geometric description given in chapter 5, then for $\psi \in \Omega^2_{X/\mathbb{Q},x}$

$$z'(0)(\psi) = \mathrm{Res}_x(\psi \wedge \varphi), \tag{7.12}$$

where φ is given in (7.11) above.

Conclusion: *With the identification (7.9), we have*

$$z'(0) = \varphi.$$

The proof of the claim is by direct computation. For example, if

$$\begin{aligned} z(t) = & \left(a_1 t^{1/2} + a_2 t + \cdots, b_1 t^{1/2} + b_2 t + \cdots\right) \\ & + \left(-a_1 t^{1/2} + a_2 t + \cdots, -b_1 t^{1/2} + b_2 t + \cdots\right) \end{aligned}$$

then

$$\varphi = \frac{1}{\xi\eta}\left[\left(2a_2 + \frac{a_1^2}{\xi} + \frac{a_1 b_1}{\eta}\right) d\eta - \left(2b_2 + \frac{a_1 b_1}{\xi} + \frac{b_1^2}{\eta}\right) d\xi + (b_1 da_1 - a_1 db_1)\right]$$

and using (7.12)

$$\begin{aligned} z'(0)(d\xi \wedge d\eta) &= b_1 da_1 - a_1 db_1 \\ z'(0)(d\xi \wedge d\alpha) &= -2a_2 d\alpha \\ z'(0)(d\eta \wedge d\alpha) &= -2b_2 d\alpha \\ z'(0)(\eta d\eta \wedge d\alpha) &= -b_1^2 d\alpha \\ z'(0)(\xi d\eta \wedge d\alpha) &= z'(0)(\eta d\xi \wedge d\alpha) = -a_1 b_1 \\ z'(0)(\xi d\xi \wedge d\alpha) &= -a_1^2 . \end{aligned}$$

Finally, regarding the canonical element ω given by (7.10), we have that

$$\omega \in \mathbb{H}^4_x\left(\Omega^\bullet_{X/\mathbb{C}}\right) \cong \mathbb{C}$$

is a generator, which topologically may be thought of as the local fundamental class of the point x.

Chapter Eight

Tangents to Related Spaces

In the preceding chapter we have given the formal definition

$$\underline{\underline{T}}Z^2(X) = \lim_{\left\{\substack{Z \text{ codim } 2 \\ \text{subscheme}}\right.} \mathcal{E}xt^2_{\mathcal{O}_X}\left(\mathcal{O}_Z, \Omega^1_{X/\mathbb{Q}}\right) \tag{8.1}$$

of the tangent sheaf to the space $Z^2(X)$ of 0-cycles on a smooth algebraic surface X. For the study of the infinitesimal geometry of $Z^2(X)$—especially the subspace $Z^2_{\text{rat}}(X)$ of 0-cycles that are rationally equivalent to zero—it is important to define and study the tangent sheaf to several related spaces. This is the objective of the first two sections of this chapter, and in the last section we will relate these to $\underline{\underline{T}}Z^2(X)$ and arrive at the main result of this work.

In Appendix A to the first section we will recall the construction and some properties of the Cousin flasque resolution of a coherent sheaf on X, and in Appendix B we will use this material in dualized form to give a description of $\underline{\underline{T}}Z^1(X)$ in terms of differential forms.

8.1 DEFINITION OF $\underline{\underline{T}}Z^1(X)$ FOR A SURFACE X

For a Y a smooth curve, based upon heuristic geometric considerations in Chapter 2 we gave the provisional definition

$$\underline{\underline{T}}Z^1(Y) = \mathcal{PP}_Y \tag{8.2}$$

for the tangent sheaf to the space of divisors on Y. In this case, divisors are the same as 0-cycles; in Chapter 2 and again in the preceding chapter we have noted that the analogue of the formal definition (8.1) when $n = 1$ agrees with (8.2).

We begin by noting that the heuristic reasoning leading to (8.2) for curves works equally well for divisors in all dimensions: For any smooth variety X and point x, given $f, g \in \mathcal{O}_{X,x}$ we have that one reasonable prescription is

$$\text{tangent at } t = 0 \text{ and localized at } x \text{ corresponds to } \operatorname{div}(f + tg) = [g/f]_x$$

where the right-hand side is the principal part at x of g/f. This suggests that for the same geometric reason as in the case of curves we take as provisional definition

$$\underline{\underline{T}}Z^1(X) = \mathcal{PP}_X,$$

where as usual the sheaf of principal parts is

$$\mathcal{PP}_X = \underline{\underline{\mathbb{C}}}(X)/\mathcal{O}_X. \tag{8.3}$$

We will now show that for X a surface[1]

$$(8.4) \qquad \mathcal{PP}_X \cong \lim_{\left\{\substack{Z \text{ codim } 1 \\ \text{subscheme}}\right.} \mathcal{E}xt^1_{\mathcal{O}_X}(\mathcal{O}_Z, \mathcal{O}_X).$$

This then leads to the formal

Definition: *The tangent sheaf to the space of codimension* 1 *cycles on a smooth variety X is*

$$\underline{\underline{T}}Z^1(X) = \lim_{\left\{\substack{Z \text{ codim } 1 \\ \text{subscheme}}\right.} \mathcal{E}xt^1_{\mathcal{O}_X}(\mathcal{O}_Z, \mathcal{O}_X).$$

Proof of (8.4): From the Appendix to this section we recall the Cousin flasque resolution (the notations are explained there)

$$0 \to \mathcal{O}_X \to \underline{\underline{\mathbb{C}}}(X) \to \bigoplus_{\left\{\substack{Y \text{ codim } 1 \\ Y \text{ irred}}\right.} \underline{\underline{H}}^1_y(\mathcal{O}_X) \to \bigoplus_{x \in X} \underline{\underline{H}}^2_x(\mathcal{O}_X) \to 0.$$

Thus we have to show that

$$(8.5) \qquad \lim_{\left\{\substack{Z \text{ codim } 1 \\ \text{subscheme}}\right.} \mathcal{E}xt^1_{\mathcal{O}_X}(\mathcal{O}_Z, \mathcal{O}_X) \cong \ker\left\{ \bigoplus_{\left\{\substack{Y \text{ codim } 1 \\ Y \text{ irred}}\right.} \underline{\underline{H}}^1_y(\mathcal{O}_X) \to \bigoplus_{x \in X} \underline{\underline{H}}^2_x(\mathcal{O}_X) \right\}.$$

This is also proved in the Appendix. Briefly the idea is this: Working in the stalk at $x \in X$ we consider an element

$$g/f \in \underline{\underline{\mathbb{C}}}(X)_x,$$

where $f, g \in \mathcal{O}_{X,x}$ are relatively prime. This gives the element

$$(8.6) \qquad \begin{cases} 0 \to F_1 \xrightarrow{f} F_0 \to \mathcal{O}_Z \to 0 & \qquad (F_1, F_0 \cong \mathcal{O}_X), \\ F_1 \xrightarrow{g} \mathcal{O}_X \end{cases}$$

in $\mathcal{E}xt^1_{\mathcal{O}_X}(\mathcal{O}_Z, \mathcal{O}_X)_x$, where the top row is the free resolution of the codimension-1 subscheme Z defined by (f) and the second row gives an element of $\mathcal{E}xt^1$, defined as usual to be the first derived functor of $\mathrm{Hom}_{\mathcal{O}_X}(\mathcal{O}_Z, \mathcal{O}_X)$. This prescription annihilates $\mathcal{O}_{X,x} \subset \underline{\underline{\mathbb{C}}}(X)_x$ and behaves in a compatible way if we have an inclusion $Z \subseteq Z'$ of subschemes. If

$$f = f_1^{l_1} \cdots f_k^{l_k},$$

where the $f_i \in \mathcal{O}_{X,x}$ are irreducible, relatively prime, and with divisor Y_i, then we map (8.6) to

$$\bigoplus_{\left\{\substack{Y \text{ codim } 1 \\ Y \text{ irred}}\right.} \underline{\underline{H}}^1_y(\mathcal{O}_X)_x$$

[1] This result is true for all dimensions; we will only use it for surfaces.

by sending (8.6) to $\sum_i \underline{\underline{H}}^1_{y_i}(\mathcal{O}_X)_x$, where the ith component is

$$\begin{cases} F_1 \xrightarrow{f_i^{l_i}} F_0 \to \mathcal{O}_X/\mathcal{I}_{Y_i}^{l_i} & (F_0, F_1 \cong \mathcal{O}_X), \\ F_1 \xrightarrow{g/f_1^{l_1}..\hat{f}_i^{l_i}..f_k^{l_k}} \mathcal{O}_X. \end{cases}$$

This maps to zero in $H_x^2(\mathcal{O}_X)$ and by exactness of the Cousin flasque resolution has kernel $\underline{\underline{\mathbb{C}}}(X)_x$ (cf. the Appendix). □

We thus have that for a surface X

$$\underline{\underline{T}}Z^1(X) \cong \ker \left\{ \bigoplus_{\left\{\substack{Y \text{ codim } 1 \\ Y \text{ irred}}\right.} \underline{\underline{H}}^1_y(\mathcal{O}_X) \to \bigoplus_{x \in X} \underline{\underline{H}}^2_x(\mathcal{O}_X) \right\} \tag{8.7}$$

whereas for a smooth curve Y we simply have

$$\underline{\underline{T}}Z^1(Y) \cong \bigoplus_{y \in Y} \underline{\underline{H}}^1_y(\mathcal{O}_Y),$$

and one question is:

What is the geometric meaning of being in the kernel in (8.7)?

Later in this chapter we will discuss this using the interpretation of the last two terms in the Cousin flasque resolution by means of differential forms, which gives a particularly geometric way of understanding the situation. Here we shall give a more algebraic explanation.

The simplest interesting example is if

$$\begin{cases} Y_1 = \text{div } f_1, Y_2 = \text{div } f_2, \\ Y_1 \cap Y_2 = x, \end{cases}$$

where the intersection is transverse

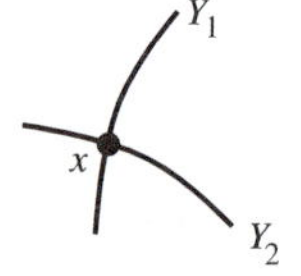

If we consider $g/f_1f_2 \in \mathbb{C}(X)_x$, then there exist $h_1, h_2 \in \mathcal{O}_{X,x}$ such that

$$g = h_1f_1 + h_2f_2 \Leftrightarrow g(x) = 0.$$

Such g's correspond to deforming Y_1 and Y_2 independently. The more interesting case when $g(x) \neq 0$ corresponds to smoothing the singularity and deforming $Y_1 + Y_2$ into an irreducible curve (all of this is local)

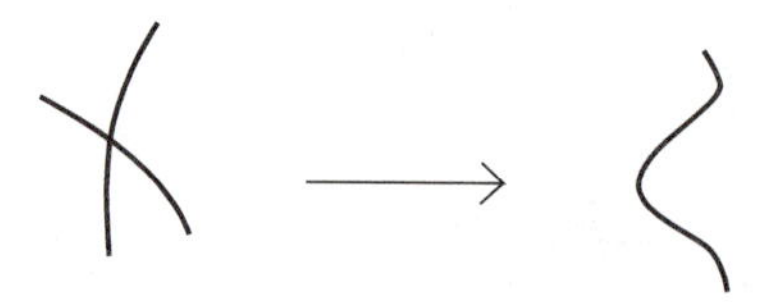

In this case, on Y_1 we have the element

$$\begin{cases} F_1 \xrightarrow{f_1} F_0 \to \mathcal{O}_{Y_1} & (F_1, F_0 \cong \mathcal{O}_X), \\ F_1 \xrightarrow{g/f_2} \mathcal{O}_X \end{cases}$$

in $H^1_{y_1}(\mathcal{O}_X)$ (the reason that poles are allowed in the second term is explained in the Appendix), and in $H^1_{y_2}(\mathcal{O}_X)$ the element

$$\begin{cases} F_1 \xrightarrow{f_2} F_0 \to \mathcal{O}_{Y_2} & (F_1, F_0 \cong \mathcal{O}_X), \\ F_1 \xrightarrow{g/f_1} \mathcal{O}_X. \end{cases}$$

The first element maps in $H^2_x(\mathcal{O}_X)$ to

$$\begin{cases} F_2 \xrightarrow{\binom{f_2}{-f_1}} F_2 \xrightarrow{(f_1, f_2)} F_0 \to \mathcal{O}_x \to 0, & (F_2, F_0 \cong \mathcal{O}_X, F_1 \cong \mathcal{O}_X \oplus \mathcal{O}_X), \\ F_2 \xrightarrow{g} \mathcal{O}_X \end{cases}$$

where the top row is the Koszul resolution of $\mathcal{O}_x$ with $x = \{f_1 = f_2 = 0\}$. The second element maps similarly, and because of the commutative diagram

$$\begin{array}{ccccccccc} 0 \to \mathcal{O}_X & \xrightarrow{\binom{f_2}{-f_1}} & \mathcal{O}_X^2 & \xrightarrow{(f_1, f_2)} & \mathcal{O}_X & \to & \mathcal{O}_x & \to & 0 \\ \downarrow -1 & & \downarrow \begin{pmatrix} 0 & 1 \\ 1 & 0 \end{pmatrix} & & \downarrow 1 & & \| & & \\ 0 \to \mathcal{O}_X & \xrightarrow{\binom{f_1}{-f_2}} & \mathcal{O}_X^2 & \xrightarrow{(f_2, f_1)} & \mathcal{O}_X & \to & \mathcal{O}_x & \to & 0 \end{array}$$

we see that these determine classes in

$$\mathcal{E}xt^2_{\mathcal{O}_X}(\mathcal{O}_x, \mathcal{O}_X)$$

that are negatives of each other and hence cancel. We thus map to 0 in $H^2_x(\mathcal{O}_X)$, as must be the case.

More sophisticated examples that illustrate the various aspects of the geometric significance of being in the kernel in (8.7) may be especially easily computed using differential forms; this will be done in section 8.2.

The simplest example of a tangent vector to a codimension-1 cycle is given by a normal vector field

$$\nu \in H^0(N_{Y/X})$$

to a smooth curve $Y \subset X$. If locally Y is the divisor of a function f, then

$$\nu = g\, \partial/\partial f$$

for some function g. (This expression is well-defined as a normal vector field along Y — its value on a 1-form φ is given by

$$\nu \rfloor \varphi = \operatorname{Res}_Y \left(\frac{g\varphi \wedge df}{f} \right)$$

where Res_Y is the Poincaré residue.) The corresponding element of $\mathcal{E}xt^1$ is given by

$$\begin{cases} F_1 \xrightarrow{\ f\ } F_0 \to \mathcal{O}_Y, \\ F_1 \xrightarrow{\ g\ } \mathcal{O}_X. \end{cases}$$

One interesting point that arises when we consider $TZ^1(X)$ rather than $T\,\mathrm{Hilb}^1(X)$ is the following question:

(8.8) *Can ν be extended to a second order arc in $Z^1(X)$?*

It is well known that this may not be possible in $\mathrm{Hilb}^1(X)$—i.e., ν *may be obstructed.* In [39] Ting Fei Ng has proved that if we allow Y to deform *as a* 1*-cycle* then this obstruction vanishes, so that (8.8) has an affirmative answer. In fact, Ng has shown that $Z^1(X)$ *is formally unobstructed*; i.e., any $\tau \in TZ^1(X)$ is tangent to an infinite order arc in $Z^1(X)$ (this is always true locally). The heuristic reason why (8.8) should be true will be discussed in section 10.

Finally, the differential of the Abel-Jacobi map

$$Z^1(X) \to \mathrm{Pic}(X)$$

is a map

$$TZ^1(X) \xrightarrow{dAJ^1_X} H^1(\mathcal{O}_X)$$

that may be expressed using Serre duality and the pairing

$$\left(\lim_{\left\{ \substack{Z \text{ codim } 1 \\ \text{subscheme}} \right.} \mathcal{E}xt^1_{\mathcal{O}_X}(\mathcal{O}_Z, \mathcal{O}_X) \right) \otimes H^1\left(\Omega^2_{X/\mathbb{C}}\right) \to \mathbb{C}$$

induced by globalizing the pairing

$$\mathcal{E}xt^1_{\mathcal{O}_X}(\mathcal{O}_Z, \mathcal{O}_X) \otimes \Omega^2_{X/\mathbb{C}} \to \mathcal{E}xt^1_{\mathcal{O}_X}\left(\mathcal{O}_X, \Omega^2_{X/\mathbb{C}}\right) \cong \omega_Z$$

and composing with the trace mapping

$$H^1(\omega_Z) \xrightarrow{\mathrm{Tr}} \mathbb{C}.$$

8.1.1 The Cousin flasque resolution; duality (cf. [13], [14], and [15])

Let X be a smooth n-dimensional quasi-projective algebraic variety and denote by $V^p(X)$ the set of all irreducible, codimension-p subvarieties of X. The Cousin flasque resolution for the sheaf $\mathcal{O}_X$ is

$$\text{(A.1)} \qquad 0 \to \mathcal{O}_X \to \underline{\underline{\mathbb{C}}}(X) \to \bigoplus_{Y \in V^1(X)} \underline{\underline{H}}^1_y(\mathcal{O}_X) \to \bigoplus_{Z \in V^2(X)} \underline{\underline{H}}^2_z(\mathcal{O}_X) \to \cdots$$

$$\cdots \to \bigoplus_{x \in X} \underline{\underline{H}}^n_x(\mathcal{O}_X) \to 0.$$

Here, $y \in Y$, $z \in Z \ldots$, are generic points, $\underline{\underline{H}}^1_y(\mathcal{O}_X)$, $\underline{\underline{H}}^2_z(\mathcal{O}_X), \ldots$ denote the Zariski sheaves $(j_Y)_* H^1_y(\mathcal{O}_X)$, $(j_Z)_* H^2_z(\mathcal{O}_X)$, and we identify X with $V^n(X)$. The objectives of this appendix are (i) to give procedures for calculating in practice the terms in (A.1), and (ii) to give ways of interpreting the terms in (A.1) using differential forms. This latter will be used in Appendix B to give an effective computational method for understanding the question below (8.7) above. We will do this in the cases $n = 1, 2$ as this is all that is needed for the present work (in any case, most of the essential features of the general case already appear here).

We first describe the procedure for computing the stalk at $x \in X$ of $\underline{\underline{H}}^1_y(\mathcal{O}_X)$, where Y is an irreducible divisor in X. We may assume that $x \in Y$ and let $f \in \mathcal{O}_{X,x}$ give a local defining equation for Y. We then have

$$(\mathrm{A.2})_l \qquad F_1 \xrightarrow{f^l} F_0 \to \mathcal{O}_{X,x}/f^l\mathcal{O}_{X,x},$$

where F_1, F_0 are copies of $\mathcal{O}_{X,x}$. We shall denote by

$$\mathrm{Hom}\,(F_1, \mathcal{O}_X)$$

the homomorphisms given by rational functions $h/g \in \mathbb{C}(X)$, where $h, g \in \mathcal{O}_{X,x}$ and g is relatively prime to f. The diagrams $(\mathrm{A.2})_l$ and $(\mathrm{A.2})_{l+1}$ are related by

$$\begin{array}{ccccc} F_1 & \xrightarrow{f^l} & F_0 & \longrightarrow & \mathcal{O}_{X,x}/f^l\mathcal{O}_{X,x} \\ {\scriptstyle f}\uparrow & & \uparrow{\scriptstyle 1} & & \uparrow \\ F_1' & \xrightarrow{f^{l+1}} & F_0' & \longrightarrow & \mathcal{O}_{X,x}/f^{l+1}\mathcal{O}_{X,x} \end{array}$$

which induces a map

$$\mathrm{Hom}\,(F_1, \mathcal{O}_X)/\mathrm{Hom}\,(F_0, \mathcal{O}_X) \xrightarrow{\alpha} \mathrm{Hom}\,(F_1', \mathcal{O}_X)/\mathrm{Hom}\,(F_0', \mathcal{O}_X).$$

If we identify $h/g \in \mathrm{Hom}\,(F_1, \mathcal{O}_X)$ with the function h/gf^l, then $\alpha(h/g) = h'/g'$ where $h' = fh$, $g' = g$, then we have consistency:

$$h/gf^l = h'/g'f^{l+1}.$$

The prescription for the stalk at x of the sheaf $\underline{\underline{H}}^1_y(\mathcal{O}_X)$ is

$$(\mathrm{A.3}) \qquad \underline{\underline{H}}^1_y(\mathcal{O}_X)_x = \varinjlim_l \mathrm{Hom}\,(F_1, \mathcal{O}_X)/\mathrm{Hom}\,(F_0, \mathcal{O}_X).$$

By what was just said, there is a map

$$\mathcal{PP}_{X,x} \to \left(\bigoplus_{Y \in V^1(X)} \underline{\underline{H}}^1_y(\mathcal{O}_X) \right)_x$$

as follows: Write $f \in \mathbb{C}(X)$ as

$$f = h/f_1^{l_1} \cdots f_k^{l_k}$$

where $f_1, \cdots, f_k$, $h \in \mathcal{O}_{X,x}$ are relatively prime and the f_i are irreducible. Set $Y_i = \{f_i = 0\}$. Then

$$f \to \sum_i h/f_1^{l_1} \cdots \hat{f}_i^{l_i} \cdots f_k^{l_k}$$

where the $\hat{f}_i^{l_i}$ means to omit the ith term and $h/f_1^{l_1}\cdots\hat{f}_i^{l_i}\cdots f_k^{l_k}$ is considered as an element of $\underline{\underline{H}}^1_{y_i}(\mathcal{O}_X)_x$ by the procedure given above.

When $n=1$ we obviously have

$$\mathcal{PP}_{X,x}\cong H^1_x(\mathcal{O}_X).$$

When $n=2$ the map

$$\underline{\underline{\mathbb{C}}}(X)_x\to\left(\bigoplus_{Y\in V^1(X)}\underline{\underline{H}}^1_y(\mathcal{O}_X)\right)_x$$

clearly has kernel $\mathcal{O}_{X,x}$, and hence it gives an injection

$$\mathcal{PP}_{X,x}\hookrightarrow\left(\bigoplus_{Y\in V^1(X)}\underline{\underline{H}}^1_y(\mathcal{O}_X)\right)_x.$$

We shall give the prescription for computing $H^2_x(\mathcal{O}_X)$ and then observe that from the general theory it follows that the exact sequence

$$\mathcal{PP}_{X,x}\to\left(\bigoplus_{Y\in V^1(X)}\underline{\underline{H}}^1_y(\mathcal{O}_X)\right)_x\to H^2_x(\mathcal{O}_X)\to 0 \tag{A.4}$$

is exact.

For X a surface and $Z\subset X$ any subscheme with supp $Z=x$ and ideal $\mathcal{I}_Z\subset\mathcal{O}_{X,x}$, we take a minimal resolution

$$0\to E_2\to E_1\to E_0\to\mathcal{O}_Z\to 0$$

where $E_0\cong\mathcal{O}_{X,x}$ and E_i is isomorphic to the direct sum of $\mathcal{O}_{X,x}$'s for $i=1,2$. We then consider the quotient

$$\text{Hom}\left(E_2,\mathcal{O}_{X,x}\right)/\text{Hom}\left(E_1,\mathcal{O}_{X,x}\right)$$

where the matrix elements in Hom $(E_i,\mathcal{O}_{X,x})$ are in $\mathcal{O}_{X,x}$. If Z' is another subscheme with $\mathcal{I}_{Z'}\subset\mathcal{I}_Z$, then there is a commutative diagram

$$\begin{array}{ccccccccc}0&\to&E_2&\to&E_1&\to&E_0&\to&\mathcal{O}_Z&\to 0\\ &&\uparrow&&\uparrow&&\uparrow&&\uparrow&\\ 0&\to&E_2'&\to&E_1'&\to&E_0'&\to&\mathcal{O}_{Z'}&\to 0\end{array}$$

inducing a map

$$\text{Hom}\left(E_2,\mathcal{O}_{X,x}\right)/\text{Hom}\left(E_1,\mathcal{O}_{X,x}\right)\to\text{Hom}\left(E_2',\mathcal{O}_{X,x}\right)/\text{Hom}\left(E_1',\mathcal{O}_{X,x}\right).$$

The prescription is

$$H^2_x(\mathcal{O}_X)=\lim_{\left\{\substack{\text{subschemes } Z\\ \text{supp } Z=x}\right.}\text{Hom}\,(E_2,\mathcal{O}_{X,x})/\text{Hom}\,(E_1,\mathcal{O}_{X,x}). \tag{A.5}$$

It is well known that we may take the limit only over the subschemes with ideal $\mathfrak{m}_x^l$ and obtain the same result.

If X is a curve and ξ is a local uniformizing parameter centered at x, then we have

$$\mathcal{PP}_{X,x} \cong H^1_x(\mathcal{O}_X) \cong \left\{ \begin{array}{l} \text{finite Laurent tails} \\ \sum_{k>0} a_k/\xi^k \end{array} \right\}.$$

For X a surface with local uniformizing parameters ξ, η centered at x, we have similarly (cf. (f)) below

$$H^2_x(\mathcal{O}_X) \cong \left\{ \begin{array}{l} \text{finite 4}^{\text{th}} \text{ quadrant Laurent} \\ \text{tails } \tau = \sum_{k,l>0} a_{kl}/\xi^k \eta^l \end{array} \right\}.$$

The map

$$\left(\bigoplus_{Y \in V^1(X)} \underline{\underline{H}}^1_y(\mathcal{O}_X) \right)_x \to H^2_x(\mathcal{O}_X)$$

may be described as follows: Represent an element of $H^1_y(\mathcal{O}_Y)$ by

$$\left\{ \begin{array}{l} F_1 \xrightarrow{f^k} F_0 \to \mathcal{O}_{X,x}/f^k\mathcal{O}_{X,x}, \\ \quad h/g^l \text{ where } h, g \in \mathcal{O}_{X,x} \end{array} \right.$$

and we assume for the moment that f and g are relatively prime and irreducible. Then f^k and g^l generate an ideal $\mathcal{I}_Z$ with supp $Z = x$, and we may consider the corresponding Koszul complex

$$0 \to E_2 \xrightarrow{\binom{g^l}{-f^k}} E_1 \xrightarrow{(f^k, g^l)} E_0 \to \mathcal{O}_Z \to 0,$$

where $E_2 \cong \mathcal{O}_{X,x}$, $E_1 \cong \mathcal{O}_{X,x} \oplus \mathcal{O}_{X,x}$, $E_0 \cong \mathcal{O}_{X,x}$, and the map $E_2 \to E_1$ is given by $1 \to (-g^l, f^k)$. Then the above element of $H^1_y(\mathcal{O}_X)$ maps to

$$h \in \operatorname{Hom}(E_2, \mathcal{O}_{X,x})/\operatorname{Hom}(E_1, \mathcal{O}_{X,x}).$$

In general, h/g^l will be of the form $h/g_1^{l_1} \cdots g_k^{l_k}$ where $f, g_1, \ldots, g_k$ are irreducible and pairwise relatively prime. We then sum up the above construction with $(f^k, g_i^{l_i})$ replacing (f^k, g^l).

By our sign conventions, the sequence (A.4) is a complex and it is a nontrivial result that it is exact. The surjectivity on the right is easy to check using the Laurent series interpretation of $H^2_x(\mathcal{O}_X)$. The issue is to show that given the data

$$\left\{ \begin{array}{l} F_{1,i} \xrightarrow{f_i^{l_i}} F_{0,i} \to \mathcal{O}_{X,x}/f_i^{l_i}\mathcal{O}_{X,x}, \\ h_i/g_i \in \operatorname{Hom}(F_{1,i}, \mathcal{O}_{X,x})/\operatorname{Hom}(F_{0,i}, \mathcal{O}_{X,x}), \end{array} \right.$$

where $1 \leqq i \leqq k$ and the $Y_i = \{f_i = 0\}$ are irreducible curves passing through x, then the condition that the sum of these data in $\bigoplus_{Y_i} H^1_{y_i}(\mathcal{O}_X)$ map to zero in $H^2_x(\mathcal{O}_X)$ gives the compatibility condition that the (equivalence classes of) the h_i/g_i come from a single rational function in $\mathbb{C}(X)$.

We conclude this section by drawing global consequences of the relation between the Cousin flasque resolution

$$0 \to \mathcal{O}_X \to \underline{\underline{\mathbb{C}}}(X) \to \bigoplus_{\begin{cases} Y \text{ irred} \\ Y \text{ codim } 1 \end{cases}} \underline{\underline{H}}_y^1(\mathcal{O}_X) \to \bigoplus_{x \in X} \underline{\underline{H}}_x^2(\mathcal{O}_X) \to 0$$

and the global tangent space

$$TZ^1(X) =: H^0(\underline{\underline{T}}Z^1(X))$$

as defined above. Namely, from the acyclicity of the Zariski sheaves in the above resolution and standard identification

$$H^1(\mathcal{O}_X) \cong T\mathrm{Pic}(X)$$

we have the exact sequence

$$0 \to \mathbb{C}(X)/\mathbb{C} \to TZ^1(X) \to T\mathrm{Pic}(X) \to 0.$$

The geometric interpretation of this is clear.

Before turning to the interpretation of the terms in the Cousin flasque resolution via differential forms, we want to give a

Summary of Grothendieck local duality: Let X be an n-dimensional smooth variety and Z a subscheme with supp $Z = x$. Then

(a) There is a natural map

$$\mathcal{E}xt^n_{\mathcal{O}_X}(\mathcal{O}_Z, \Omega^n_{X/\mathbb{C}}) \xrightarrow{\mathrm{Tr}} \mathbb{C}.$$

(b) The $\mathcal{O}_X$-modules $\mathcal{E}xt^n_{\mathcal{O}_X}(\mathcal{O}_Z, \mathcal{O}_X)$ and $\mathcal{E}xt^n_{\mathcal{O}_X}(\mathcal{O}_Z, \Omega^n_{X/\mathbb{C}})$ annihilate $\mathcal{I}_Z$; i.e., they are $\mathcal{O}_Z$-modules.

(c) The pairing

$$\text{(A.6)} \qquad \begin{array}{ccc} \mathcal{O}_Z \bigotimes \mathcal{E}xt^n_{\mathcal{O}_X}(\mathcal{O}_Z, \Omega^n_{X/\mathbb{C}}) & \longrightarrow & \mathcal{E}xt^n(\mathcal{O}_Z, \Omega^n_{X/\mathbb{C}}) \\ & \searrow & \downarrow \mathrm{Tr} \\ & & \mathbb{C} \end{array}$$

is a perfect pairing (local duality).

(d) Given a regular sequence $f_1, \ldots, f_n$ with $\mathcal{I}_Z = (f_1, \ldots, f_n)$, there is an identification $\mathcal{E}xt^n_{\mathcal{O}_X}(\mathcal{O}_Z, \Omega^n_{X/\mathbb{C}}) \cong \Omega^n_{X/\mathbb{C},x} \otimes \mathcal{O}_Z$ and the pairing

$$\mathcal{O}_Z \otimes \left(\Omega^n_{X/\mathbb{C},x} \otimes \mathcal{O}_Z\right) \to \mathbb{C}$$

is given by

$$g \otimes (\omega \otimes h) \to \mathrm{Res}_x \left\{ \frac{gh\omega}{f_1 \cdots f_n} \right\}$$

where the term in brackets is the Grothendieck residue symbol.

(e) In general, given a minimal free resolution

$$0 \to E_n \to E_{n-1} \to \cdots \to E_0 \to \mathcal{O}_Z \to 0$$

there is a recipe for computing the pairing (A.6) in terms of Grothendieck residue symbols.

(f) Local duality for subschemes Z with $\operatorname{supp} Z = x$ together with the isomorphism

$$\lim_{\left\{\substack{Z \text{ codim } n \\ \text{subscheme}}\right.} \mathcal{E}xt^n_{\mathcal{O}_X}(\mathcal{O}_Z, \mathcal{O}_X) \cong H^n_x(\mathcal{O}_X)$$

give an isomorphism

(A.7) $$H^n_x(\mathcal{O}_X) \cong \operatorname{Hom}^c_{\mathbb{C}}\left(\Omega^n_{X/\mathbb{C},x}, \mathbb{C}\right).$$

When $n = 2$, if we make the identification of $H^2_x(\mathcal{O}_X)$ with fourth quadrant Laurent tails τ as above, then (A.7) is given by

$$\tau(\omega) = \operatorname{Res}(\tau\omega),$$

where the residue is the usual interated 1-variable residue.

Proof: If ξ, η are local coordinates at x, and ξ^*, η^* are the dual variables, then we may naturally identify

$$\mathcal{E}xt^2_{\mathcal{O}_X}\left(\mathcal{O}_X/\mathfrak{m}_x^{k+1}, \mathcal{O}_X\right) = \left\{ \sum_{\substack{i+j \le k \\ i,j \ge 0}} a_{ij}\xi^{*i}\eta^{*j} \right\}.$$

To see this, we have the minimal free resolution

$$0 \to \mathcal{O}_X^{k+1} \xrightarrow{\begin{pmatrix} \eta & 0 & \cdots & 0 \\ -\xi & \eta & \cdots & 0 \\ 0 & -\xi & \cdots & 0 \\ \cdot & \cdot & & \cdot \\ \cdot & \cdot & & \cdot \\ \cdot & \cdot & & \eta \\ 0 & 0 & \cdots & -\xi \end{pmatrix}} \mathcal{O}_X^{k+2} \xrightarrow{(\xi^{k+1}, \xi^k\eta, \cdots, \eta^{k+1})} \mathcal{O}_X$$

$$\to \mathcal{O}_X/\mathfrak{m}_x^{k+1} \to 0.$$

If we let $e_1, \ldots, e_{k+1}$ be a basis for $\mathcal{O}_X^{k+1}$, then if we identify

$$e_1^* \leftrightarrow \xi^{*k}, e_2^* \leftrightarrow \xi^{*k-1}\eta^*, \ldots, e_{k+1}^* \leftrightarrow \eta^{*k}$$

then the relations in $\mathcal{E}xt^2_{\mathcal{O}_X}\left(\mathcal{O}_X/\mathfrak{m}_x^{k+1}, \mathcal{O}_X\right)$

$$\eta e_2^* = \xi e_1^*$$

$$\vdots$$

$$\eta e_{k+1}^* = \xi e_k^*$$

$$0 = \xi e_{k+1}^*$$

translate into the relations

$$\begin{aligned}\eta \rfloor \xi^{*k} &= 0\\ \eta \rfloor \xi^{*k-1}\eta^{*} &= \xi \rfloor \xi^{*k}\\ &\vdots\\ \eta \rfloor \eta^{*k} &= \xi \rfloor \xi^{*}\eta^{*k-1}\\ 0 &= \xi \rfloor \eta^{*k}.\end{aligned}$$

Consequently

$$\begin{aligned}\mathcal{E}xt^2_{\mathcal{O}_X}\left(\mathcal{O}_X/\mathfrak{m}_x^{k+1}, \mathcal{O}_X\right) &= \mathrm{Hom}_{\mathcal{O}_X}\left(\mathcal{O}_X^{k+1}, \mathcal{O}_X\right)/\mathrm{Hom}_{\mathcal{O}_X}\left(\mathcal{O}_X^{k+2}, \mathcal{O}_X\right)\\ &\cong \left\{\sum_{l=0}^{k} p_l(\xi,\eta)\rfloor \xi^{*k-l}\eta^{*l} : p_l(\xi,\eta) \in \mathcal{O}_X\right\}\\ &= \left\{\sum_{\substack{i+j\le k\\ i,j\ge 0}} a_{ij}\xi^{*i}\eta^{*j}\right\}.\end{aligned}$$

Here $\rfloor$ means

$$\xi^a\eta^b \rfloor \xi^{*i}\eta^{*j} = \begin{cases}\xi^{*i-a}\eta^{*j-b} & \text{if } a \le i, b \le j,\\ 0 & \text{otherwise.}\end{cases}$$

Under the above isomorphism, Grothendieck local duality is the nondegenerate pairing

$$\mathcal{E}xt^2_{\mathcal{O}_X}\left(\mathcal{O}_X/\mathfrak{m}_x^{k+1}, \mathcal{O}_X\right) \otimes \left(\Omega^2_{X/\mathbb{C}} \otimes \mathcal{O}_X/\mathfrak{m}_x^{k+1}\right) \to \mathbb{C},$$

where

$$\sum_{\substack{i+j\le k\\ i,j\ge 0}} a_{ij}\xi^{*i}\eta^{*j} \otimes \sum_{\substack{i+j\le k\\ i,j\ge 0}} b_{ij}\xi^{i}\eta^{j}\, d\xi \wedge d\eta \mapsto \sum_{\substack{i+j\le k\\ i,j\ge 0}} a_{ij}b_{ij}.$$

If we make the identification with Laurent tails

$$\sum a_{ij}\xi^{*i}\eta^{*j} \leftrightarrow \sum a_{ij}\xi^{-i-1}\eta^{-j-1}$$

then the duality becomes

$$\tau \otimes \omega \mapsto \mathrm{Res}_x(\tau\omega)$$

since $\xi^{*i}\eta^{*j} \leftrightarrow \frac{1}{\xi^{i+1}\eta^{j+1}}$ under the residue mapping.

If we pass to the limit,

$$\begin{aligned}H^2_x(\mathcal{O}_X) &= \left\{\sum_{\left\{\substack{i,j\geqq 0\\ \text{finite sum}}\right.} a_{ij}\xi^{*i}\eta^{*j}\right\}\\ &\cong \left\{\sum_{\left\{\substack{i,j<0\\ \text{finite sum}}\right.} b_{ij}\xi^{i}\eta^{j}\right\}\end{aligned}$$

and Grothendieck local duality is the nondegenerate pairing

$$H^2_x(\mathcal{O}_X) \otimes \Omega^2_{X/\mathbb{C},x} \to \mathbb{C},$$
$$\tau \otimes \omega \mapsto \operatorname{Res}_x(\tau\omega).$$

Thus

$$H^2_x(\mathcal{O}_X) \cong \{\text{4th quadrant finite Laurent tails at } x\}$$

with Grothendieck duality being given by residue.

We note that

$$H^2_x\left(\Omega^1_{X/\mathbb{Q}}\right) = \left\{ \sum_{\left\{\substack{i,j<0 \\ \text{finite sum}}\right.} \alpha_{ij}\xi^i\eta^j : \alpha_{ij} \in \Omega^1_{X/\mathbb{Q}}\Big|_x \right\}.$$

In conclusion,

for X a surface we have a natural isomorphism

$$H^2_x(\mathcal{O}_X) \cong \operatorname{Hom}^c_{\mathbb{C}}\left(\Omega^2_{X/\mathbb{C},x}, \mathbb{C}\right).$$

This expresses the last term in the Cousin flasque resolution

$$0 \to \mathcal{O}_X \to \underline{\underline{\mathbb{C}}}(X) \to \underset{Y}{\oplus} H^1_y(\mathcal{O}_X) \to \underset{x}{\oplus} H^2_x(\mathcal{O}_X) \to 0$$

in terms of differentials.

It remains to so express the third term, to which we will turn after a side discussion of the local arithmetic fundamental class.

Local arithmetic cycle class of a 0-dimensional subscheme: Angéniol and Lejeune-Jalabert [19] have given a definition—referred to in section 7.2 above—of the local fundamental class

$$[Z]_{\text{loc}} \in H^n_x\left(\Omega^n_{X/\mathbb{C}}\right)$$

of a 0-dimensional subscheme Z with support $Z = x$. Recall that if

$$0 \to F_n \xrightarrow{f_n} F_{n-1} \to \cdots \to F_1 \xrightarrow{f_1} \mathcal{O}_X \to \mathcal{O}_Z \to 0$$

is a free resolution of $\mathcal{O}_Z$ where each $F_i \cong \mathcal{O}^{r_i}_X$, so that the f_i are matrices with entries in $\mathcal{O}_X$, then

$$\frac{1}{n!} df_1 \circ \cdots \circ df_n \in \operatorname{Hom}\left(F_n, \Omega^n_{X/\mathbb{C}}\right)$$

defines an element in

$$\lim_k \mathcal{E}xt^n_{\mathcal{O}_X}\left(\mathcal{O}_X/\mathcal{I}^k_Z, \Omega^n_{X/\mathbb{C}}\right) \cong \lim_k \mathcal{E}xt^n_{\mathcal{O}_X}\left(\mathcal{O}_X/\mathfrak{m}^k_x, \Omega^n_{X/\mathbb{C}}\right)$$
$$\cong H^n_x\left(\Omega^n_{X/\mathbb{C}}\right)$$

which does not depend on the choice of resolution and local trivializations. By definition this is the class $[Z]_{\text{loc}}$.

Now by (A.6)

$$H^n_x\left(\Omega^n_{X/\mathbb{C}}\right) \cong \text{dual of } \lim_k \left(\mathcal{O}_X/\mathfrak{m}^i_x\right).$$

For Z the subscheme x defined by $\mathfrak{m}_x$, using the above identification the standard Koszul resolution of $\mathfrak{m}_x$ gives the interpretation of $[x]_{\mathrm{loc}}$ as the evaluation map

$$[x]_{\mathrm{loc}}(f) = f(x), \qquad f \in \mathcal{O}_X.$$

More generally, by an argument similar to step 3 in the proof of (7.8) in section 7.2, we have for any 0-dimensional subscheme Z supported at x that

$$[Z]_{\mathrm{loc}}(f) = l(Z)f(x)$$

where $l(Z)$ is the length of Z.

Now the Angéniol and Lejeune-Jalabert construction may be adapted to define the *local arithmetic fundamental class*

$$[Z]_{a,\mathrm{loc}} \in H^n_x\left(\Omega^n_{X/\mathbb{Q}}\right)$$

as follows: Given a free resolution of $\mathcal{O}_Z$ as above, we consider

$$df_i \in \mathrm{Hom}\left(F_i, F_{i-1} \otimes \Omega^1_{X/\mathbb{Q}}\right)$$

as matrices of absolute differentials and proceed as before. Now, as explained in section 7.2 and just above,

$$\begin{aligned} H^n_x\left(\Omega^n_{X/\mathbb{Q}}\right) &\cong \lim_{\substack{\to \\ i}} \mathcal{E}xt^n_{\mathcal{O}_X}\left(\mathcal{O}_X/\mathfrak{m}^i_x, \Omega^n_{X/\mathbb{Q}}\right) \\ &\cong \lim_{\substack{\to \\ i}} \mathrm{Hom}^o_{\mathcal{O}_X}\left(\Omega^n_{X/\mathbb{Q}} \otimes \mathcal{O}_X/\mathfrak{m}^i_x, \Omega^n_{\mathbb{C}/\mathbb{Q}}\right) \end{aligned}$$

and we claim that for $\omega \in \Omega^n_{X/\mathbb{Q}}$

$$[Z]_{a,\mathrm{loc}}(\omega) = l(Z)\mathrm{ev}_x(\omega) \in \Omega^n_{\mathbb{C}/\mathbb{Q}}$$

where $\omega \to \mathrm{ev}_x(\omega)$ is the evaluation map. Explicitly,

$$\frac{1}{n!} df_1 \circ \cdots \circ df_n \wedge \omega \in \mathrm{Hom}\left(F_n, \Omega^{2n}_{X/\mathbb{Q}}\right),$$

and combining this with the natural map

$$\Omega^{2n}_{X/\mathbb{Q}} \to \Omega^n_{\mathbb{C}/\mathbb{Q}} \otimes \Omega^n_{X/\mathbb{C}}$$

we get a class

$$\left[\frac{1}{n!} df_1 \circ \cdots \circ df_n \wedge \omega\right] \in \Omega^n_{\mathbb{C}/\mathbb{Q}} \otimes \mathcal{E}xt^n_{\mathcal{O}_X}\left(\mathcal{O}_Z, \Omega^n_{X/\mathbb{C}}\right).$$

Then our claim is

$$\mathrm{Tr}\left[\frac{1}{n!} df_1 \circ \cdots \circ df_n \wedge \omega\right] = l(Z)\mathrm{ev}_x(\omega).$$

To prove this result, one first verifies it directly when $\mathcal{I}_Z = \mathfrak{m}_x$. In general, one perturbs the above resolution to get a flat family $Z(t)$ given by data $f_1(t), \ldots, f_n(t)$ and where

$$Z(t) = \sum_{i=1}^{l(Z)} x_i(t)$$

with the $x_i(t)$ distinct for $t \neq 0$. The result is true for each $x_i(t), t \neq 0$, and by taking the limit we obtain the desired statement (cf. section 7.2 for a similar argument).

For $n = 2$ the extra information in $[Z]_{a,\mathrm{loc}}$ is the difference between

$$\Omega^2_{X/\mathbb{Q}} \xrightarrow{\mathrm{ev}_x} \Omega^2_{\mathbb{C}/\mathbb{Q}}$$

and

$$\mathcal{O}_X \xrightarrow{\mathrm{ev}_x} \mathbb{C}.$$

Local duality along an irreducible curve: Let $Y \subset X$ be an irreducible curve, and let ω be a rational 2-form on X whose polar locus includes Y. We want to define a rational 1-form Y such that

$$\mathrm{Res}_Y(\omega) \in \Omega^1_{\mathbb{C}(Y)/\mathbb{C}}. \tag{A.8}$$

If ω has a first-order pole on Y, then we can take $\mathrm{Res}_Y(\omega)$ to be simply the Poincaré residue of ω along Y. (Even if ω has poles only on Y, $\mathrm{Res}_Y(\omega)$ may have singularities at the singular points of Y.) In general, to be able to define (A.8) we need to introduce auxilary data in the form of a retraction $U \longrightarrow Y$ of a Zariski open set U such that $U \cap Y$ is a Zariski open set in Y. We will not get into the formal definition here, but will just explain how the process works in practice. At a general point y of Y, we may choose rational functions $\xi, \eta \in \mathbb{C}(X)$ that are regular at y and such that $\eta = 0$ on Y. Geometrically, we have a rational mapping

$$X - - \to \mathbb{P}^2$$

which is regular near y and such that Y maps to a line. Using ξ, η as local *holomorphic* coordinates we may define $\mathrm{Res}_Y(\omega)$ in an *analytic* neighborhood of y in the usual way. Thus writing

$$\omega = \left(\frac{f_k(\xi)}{\eta^k} + \cdots + \frac{f_1(\xi)}{\eta} + f_0(\xi, \eta) \right) d\xi \wedge d\eta$$

where $f_0(\xi, \eta)$ is regular near y, we set

$$\mathrm{Res}_Y(\omega) = -f_1(\xi) d\xi.$$

We may cover a Zariski open subset Y^0 of Y with analytic neighborhoods in which this process works, and we observe that (i) the definition agrees in intersections,[2] and (ii) the resulting regular 1-form on Y^0 has at most poles on Y and therefore defines a rational 1-form on Y.

We also observe that the map $\omega \to \mathrm{Res}_Y(\omega)$ is $\mathcal{O}_Y$-linear in the following sense: At a general point $y \in Y$ the retraction gives an inclusion

$$\mathcal{O}_{Y,y} \hookrightarrow \mathcal{O}_{X,y} \tag{A.9}$$

such that the composition $\mathcal{O}_{Y,y} \to \mathcal{O}_{X,y} \xrightarrow{\text{restriction}} \mathcal{O}_{Y,y}$ is the identity. Using (A.9), a rational function f on Y induces a rational function $\widetilde{f}$ on X and

$$\mathrm{Res}_Y(\widetilde{f}\omega) = f\, \mathrm{Res}_Y(\omega).$$

[2]The point is this: We have a rational differential form $\omega = f(\xi, \eta) d\xi \wedge d\eta$ on X. At a general point of Y, locally in the analytic topology we may expand $f(\xi, \eta)$ in a Laurent series on η and define the residue as above. In the overlap of two such analytic neighborhoods we get the same answer.

Denoting by $\Omega^2_{X/\mathbb{C},Y}$ the restriction of $\Omega^2_{X/\mathbb{C}}$ to Y (thus the stalk of $\Omega^2_{X/\mathbb{C},Y}$ at x is zero if $x \notin Y$, $\Omega^2_{X/\mathbb{C},x}$ if $x \in Y$), we may define the sheaf

$$\operatorname{Hom}_{\mathcal{O}_Y}\left(\Omega^2_{X/\mathbb{C},Y}, \Omega^1_{\mathbb{C}(Y)/\mathbb{C}}\right).$$

By the preceding discussion, using the retraction we may define a sheaf mapping

$$\underline{\underline{\mathbb{C}}}(X) \to \operatorname{Hom}_{\mathcal{O}_Y}\left(\Omega^2_{X/\mathbb{C},Y}, \Omega^1_{\mathbb{C}(Y)/\mathbb{C}}\right). \tag{A.10}$$

In fact, given a retraction as above we may define a mapping

$$H^1_y(\mathcal{O}_X) \longrightarrow \operatorname{Hom}_{\mathcal{O}_Y}\left(\Omega^2_{X/\mathbb{C},Y}, \mathcal{O}^1_{\mathbb{C}(Y)}\right) \tag{A.11}$$

as follows: Given $x \in Y$ and the data

$$\begin{cases} F_1 \xrightarrow{f^l} F_0 \longrightarrow \mathcal{O}_{X,x}/f^l\mathcal{O}_{X,x} \\ F_1 \xrightarrow{g} \mathcal{O}_{X,x} \end{cases}$$

defining an element of stalk at x of the sheaf $H^1_y(\mathcal{O}_X)$, then for $\omega \in \Omega^2_{X/\mathbb{C},x}$

$$\omega \longrightarrow \operatorname{Res}_Y\left(\frac{g\omega}{f^l}\right)$$

defines the map (A.11). Moreover, (A.10) and (A.11) are compatible in the sense that the diagram

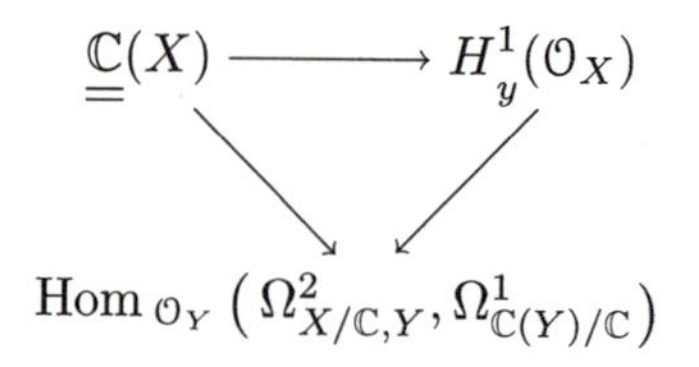

is commutative (the two slanted arrows both being defined by the same retraction). The basic fact in local duality along an irreducible curve is that *the mapping (A.11) is an isomorphism.*

We will not prove this here as our interest is in the geometric interpretations of the Cousin flasque resolution and its relation to $\underline{\underline{T}}Z^1(X)$. However, in order to complete the story we shall define, for $x \in Y$, a map

$$\operatorname{Hom}_{\mathcal{O}_Y}\left(\Omega^2_{X/\mathbb{C},x}, \Omega^1_{\mathbb{C}(Y)/\mathbb{C}}\right) \xrightarrow{\rho_x} \operatorname{Hom}^c_{\mathbb{C}}\left(\Omega^2_{X/\mathbb{C},x}, \mathbb{C}\right) \tag{A.12}$$

such that the diagram

$$\begin{array}{ccc} \bigoplus_Y \underline{\underline{H}}^1_y(\mathcal{O}_X) & \longrightarrow & \bigoplus_x \underline{\underline{H}}^2_x(\mathcal{O}_X) \\ \downarrow & & \downarrow \\ \bigoplus_Y \underline{\underline{\operatorname{Hom}}}_{\mathcal{O}_Y}\left(\Omega^2_{X/\mathbb{C},Y}, \Omega^1_{\mathbb{C}(Y)/\mathbb{C}}\right) & \longrightarrow & \bigoplus_x \underline{\underline{\operatorname{Hom}}}^c_{\mathbb{C}}\left(\Omega^2_{X/\mathbb{C},x}, \mathbb{C}\right) \end{array} \tag{A.13}$$

is commutative, where for each Y we have chosen a retraction as above. In fact, (A.12) is simply the usual residue: Given

$$\alpha \in \operatorname{Hom}_{\mathcal{O}_Y}\left(\Omega^2_{X/\mathbb{C},x}, \Omega^1_{\mathbb{C}(Y)/\mathbb{C}}\right),$$

then for $\omega \in \Omega^2_{X/\mathbb{C},x}$

$$\rho_x(\alpha)(\omega) = \mathrm{Res}_x(\alpha(\omega)),$$

where the right-hand side is the residue at $x \in Y$ of the rational 1-form $\alpha(\omega) \in \Omega^1_{\mathbb{C}(Y)/\mathbb{C}}$ (with the usual conventions if x is a singular point of Y).

In summary then, *for X a smooth surface the usual Cousin flasque resolution*

$$0 \to \mathcal{O}_X \to \underline{\underline{\mathbb{C}}}(X) \to \bigoplus_Y \underline{\underline{H}}^1_y(\mathcal{O}_X) \to \bigoplus_x \underline{\underline{H}}^2_x(\mathcal{O}_X) \to 0$$

may, upon choice of retractions, be interpreted as

$$\text{(A.14)} \qquad \begin{aligned} 0 \to \mathcal{O}_X \to \underline{\underline{\mathbb{C}}}(X) &\to \bigoplus_Y \underline{\underline{\mathrm{Hom}}}_{\mathcal{O}_Y}\left(\Omega^2_{X/\mathbb{C},Y}, \Omega^1_{\mathbb{C}(Y)/\mathbb{C}}\right) \\ &\to \bigoplus_x \underline{\underline{\mathrm{Hom}}}^c\left(\Omega^2_{X/\mathbb{C},x}, \mathbb{C}\right) \to 0. \end{aligned}$$

8.2 DUALITY AND THE DESCRIPTION OF $\underline{\underline{T}}Z^1(X)$ USING DIFFERENTIAL FORMS

In some ways the use of differential forms appropriately evaluated on tangent vectors to the space of cycles gives an especially good geometric picture. We have seen this for the case of 0-cycles, and we shall now discuss it for divisors on surfaces. Again we shall focus on the question of the geometric meaning of the kernel in (8.7).

Referring to appendix 8.1.1, upon choices of retractions for each irreducible curve Y in X there is defined a residue map

$$\text{(B.1)} \qquad \mathcal{PP}_X \xrightarrow{r} \bigoplus_Y \underline{\underline{\mathrm{Hom}}}_{\mathcal{O}_Y}\left(\Omega^2_{X/\mathbb{C},Y}, \Omega^1_{\mathbb{C}(Y)/\mathbb{C})}\right)$$

by setting for each $x \in Y \subset X$, $f \in \underline{\underline{\mathbb{C}}}(X)$, and $\omega \in \Omega^2_{X,\mathbb{C},x}$

$$\text{(B.2)} \qquad r(f)(\omega) = \bigoplus_Y \mathrm{Res}_Y(f\omega),$$

where $\mathrm{Res}_Y(f\omega) \in \underline{\underline{\Omega}}^1_{\mathbb{C}(Y),x}$. Thinking of $\mathcal{PP}_X$ as $\underline{\underline{T}}Z^1(X)$, in simplest geometric terms the map (B.1) has the following interpretation: For the special case of an irreducible curve Y one way of giving a first order deformation of Y in X is by a global section $\nu \in H^0\left(\mathrm{Hom}(\mathcal{I}_Y/\mathcal{I}^2_Y, \mathcal{O}_Y)\right)$. We may think of ν as a normal vector field to Y over a Zariski open set, and then thinking also of ν as being the tangent to a first order variation of Y we have simply that

$$\text{(B.3)} \qquad r(\nu)(\omega) = \nu \rfloor \omega\big|_Y.$$

This is the case where under the provisional definition

$$\mathcal{PP}_X = \underline{\underline{T}}Z^1(X)$$

the principal part $[f]_x \in \mathbb{C}(X)/\mathcal{O}_{X,x}$ has only a first order pole along Y and $\mathrm{Res}_Y(f\omega)$ is the usual Poincaré residue. The (nontrivial) geometric content of (B.3) is that when we identify ν with the first-order variation of Y given by $\operatorname{div}(g_0 + tg_1)$, where g_0 is a local defining equation for Y and $f = g_1/g_0$

$$\nu \rfloor \omega\big|_Y = \operatorname{Res}_Y(f\omega).$$

The case where we have an arc Z_t in $Z^1(X)$ with

$$\begin{cases} Z_0 = kY_0, \\ Z_t = Y_1(t) + \cdots + Y_k(t) \end{cases}$$

with the $Y_i(t)$ and Y_0 being irreducible curves stands in relation to the case just considered of one irreducible curve varying on the surface much as the situation of an irreducible Puiseaux series $z(t) = x_1(t) + \cdots + x_k(t)$ with $z(0) = kx$ stands in relation to a single point moving on a algebraic curve. Here, of course, higher order poles arise and for surfaces the use of retractions is necessary. The geometry underlying this is the following: In a Zariski open neighborhood U of Y_0 there will be a rational vector field that in a smaller neighborhood U^* is regular and induces a nonzero normal vector field along $Y^* = U^* \cap Y_0$. Thus U^* is foliated and the leaves give a map $U^* \to Y^*$ locally in the analytic topology. (This is generally *not* a rational map, since the integral curves of a rational vector field do not close up to algebraic curves.) Following an arc γ on Y_0 as t varies along an arc in the t-disc gives a 2-chain $\Gamma(t)$ in U^*. For $\omega \in \Omega^2_{X/\mathbb{C},x}$ assumed to be regular in U^*, we may consider the "abelian sum"

$$\int_{\Gamma(t)} \omega.$$

Taking the derivative at $t = 0$ gives an integral along γ

$$\int_\gamma \tau(\omega),$$

where $\tau(\omega) = \Omega^1_{\mathbb{C}(Y)/\mathbb{C}}$ is the value of the tangent vector

$$\tau \in \operatorname{Hom}_{\mathcal{O}_Y}\left(\Omega^2_{X/\mathbb{C},Y}, \Omega^1_{\mathbb{C}(Y)/\mathbb{C}}\right)$$

applied to ω and restricted to the above arc in $Z^1(X)$. The point here is that in this case the map

$$\underline{\underline{T}} Z^1(X) \to \bigoplus_Y \underline{\underline{\operatorname{Hom}}}_{\mathcal{O}_Y}\left(\Omega^2_{X/\mathbb{C},Y}, \Omega^1_{\mathbb{C}(Y)/\mathbb{C}}\right)$$

may be interpreted geometrically using calculus as "being basically like the case of points on a curve with dependence on an auxilary parameter."[3] Thus, in a sense, geometrically there is nothing essentially new beyond 0-cycles on a curve. (Of course the use of retractions introduces non-intrinsic data, and it is instructive to check that the various descriptions of $\underline{\underline{T}} Z^1(X)$ change in the same way when the non-intrinsic data is changed—cf. the "Afterword" to this section.)

[3]The "auxilary parameter" may be taken to be a local uniformizing parameter on Y given by the restriction to Y of a function $\eta \in \mathbb{C}(X)$. Combining this with a function $\xi \in \mathbb{C}(X)$ such that $\xi = 0$ on Y gives a retraction.

Essentially new phenomena arise when for example we have an arc Z_t in $Z^1(X)$ with

$$\begin{cases} Z_t = Y_t \text{ an irreducible curve for } t \neq 0, \\ Z_0 = Y_0' + Y_0'' \text{ is a reducible curve.} \end{cases}$$

Locally in the analytic topology, this is a special case of when we have a family of smooth curves acquiring a singularity. We shall discuss this by analyzing several examples. This will show the tangent τ always lies in

$$\text{(B.4)} \qquad \ker\left\{ \bigoplus_Y \underline{\underline{\mathrm{Hom}}}_{\mathcal{O}_Y}\left(\Omega^2_{X/\mathbb{C},Y}, \Omega^1_{\mathbb{C}(Y)/\mathbb{C}}\right) \to \bigoplus_x \underline{\underline{\mathrm{Hom}}}^c_{\mathbb{C}}\left(\Omega^2_{X/\mathbb{C},x}, \mathbb{C}\right) \right\}$$

and being in the kernel reflects limiting infinitesimal compatibility conditions (cf. the question below (8.7) above; this is what we are investigating using differential forms).

Suppose we have an arc Z_t in $Z^1(X)$ with tangent vector τ. What does it mean that τ should be in the kernel (B.4)? Locally in the analytic topology, given $x \in X$, the 1-cycle Z_0—assumed to be effective for this discussion—will have several different irreducible analytic branches Y_i passing through x, and some of the Y_i may be singular at x. Given $\omega \in \Omega^2_{X/\mathbb{C},x}$ and denoting by $\widetilde{Y}_i \to Y_i$ the normalization, the condition is

$$\text{(B.5)} \qquad \sum \mathrm{Res}_{\widetilde{Y}_{i,x}}(\tau(\omega)) = 0$$

where the sum is over all of the inverse images on all the $\widetilde{Y}_i$ of the point x. We shall first illustrate this for the two families

$$\text{(B.6a)} \qquad \xi\eta = t$$

and

$$\text{(B.6b)} \qquad \eta^2 = \xi^3 + t$$

which together with a third example given below illustrate the essential aspects of the phenomena that can arise.

The family (B.6a) may be pictured as

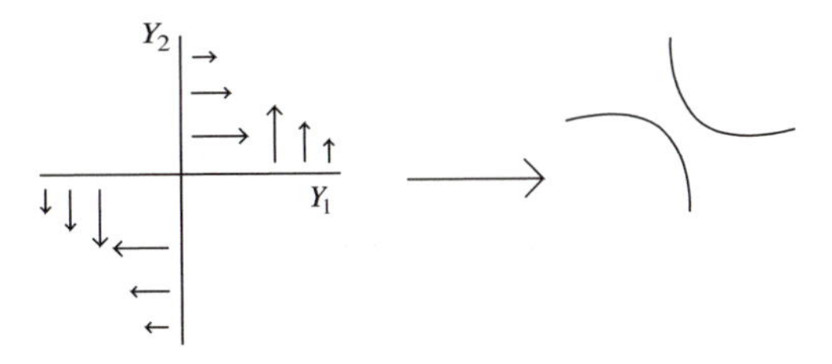

Here, Y_1 and Y_2 are the coordinate axes $\eta = 0$ and $\xi = 0$, and the arrows represent the normal vectors ν_i to the Y_i that give the tangent to the family. Explicitly

$$\nu_1 = \frac{1}{\xi}\frac{\partial}{\partial \eta},$$

$$\nu_2 = \frac{1}{\eta}\frac{\partial}{\partial \xi}.$$

By our prescription (B.3), for $\omega = f(\xi, \eta)d\xi \wedge d\eta$ we have

$$\begin{cases} \tau(\omega) \mid_{Y_1} = f(\xi, 0)\frac{d\xi}{\xi}, \\ \tau(\omega) \mid_{Y_2} = -f(0, \eta)\frac{d\eta}{\eta}, \end{cases}$$

from which (B.3) is clear. Note that the singularity acquired by Z_t produces a pole in $\tau(\omega)$. This is a geometrically distinct phenomenon from the poles produced by the singularities of the retraction when kY_0 deforms into $Y_1(t) + \cdots + Y_k(t)$ as above, when as in the curve case discussed in chapter 2 we must take the residue of a rational form with a pole of order k along Y.

For the family (B.6b), we shall use the alternate prescription discussed in earlier chapters for computing τ.[4] Namely, in $\mathbb{C}^3$ with coordinates (ξ, η, t) we consider the surface S given by (B.6b). Then for $\omega = d\xi \wedge d\eta$ we have on S

$$\begin{cases} 2d\xi = 3\eta^2 d\eta + dt, \\ \omega = \frac{1}{2}\frac{d\eta}{\xi} \wedge dt , \end{cases}$$

and this implies that

$$\tau(\omega) = \frac{1}{2}\frac{d\eta}{\xi}.$$

Uniformizing the cusp in the usual way with parameter ζ gives

$$\tau(\omega) = \frac{d\zeta}{\zeta^2}.$$

For $\omega = f(\xi, \eta)d\xi \wedge d\eta$ we have

$$\tau(\omega) = \frac{f(\zeta^3, \zeta^2)d\zeta}{\zeta^2}.$$

Thus (B.3) is satisfied.

A final interesting example is when the variation smooths a reducible curve with a multiple component; for example,

$$\xi^2\eta = t. \tag{B.6c}$$

Here, to find the answer one may replace the curve with a multiple component by one without such and take a limit

$$\xi(\xi + \lambda)\eta = t.$$

Then as above, on S and taking the limit on the component $\eta = 0$ as $\lambda \to 0$, we have

$$d\xi \wedge d\eta = \frac{d\xi \wedge dt}{\xi(\xi + \lambda)} \to \frac{d\xi}{\xi^2}.$$

The other component is more interesting: From

$$\xi^2 + \lambda\xi - \frac{1}{\eta} = 0$$

[4] cf. the Afterword below.

we have

$$\xi = \frac{\lambda \pm \sqrt{\lambda^2 - 4/\eta}}{2},$$

and thus on S

$$d\xi \wedge d\eta = \frac{\pm dt \wedge d\eta}{\eta\sqrt{\lambda^2\eta^2 - 4t}},$$

and when we sum over the two values of the square root we get zero; that is,

$$\tau(d\xi \wedge d\eta) = 0.$$

However, by a similar computation

$$\tau(\xi d\xi \wedge d\eta) = \begin{cases} -\frac{d\xi}{\xi} \text{ on } \eta = 0, \\ \frac{d\eta}{\eta} \text{ on } \xi = 0 \end{cases}$$

and again (B.3) is satisfied.

Summary: *Upon choices of retractions for each irreducible curve Y in X, and with the provisional definition* (8.2) *for* $\underline{\underline{T}}Z^1(X)$, *we have*

(B.7)

$$\underline{\underline{T}}Z^1(X) \cong \ker\left\{\bigoplus_Y \underline{\underline{\mathrm{Hom}}}_{\mathcal{O}_Y}\left(\Omega^2_{X/\mathbb{C},Y}, \Omega^1_{\mathbb{C}(Y)/\mathbb{C}}\right) \to \bigoplus_x \underline{\underline{\mathrm{Hom}}}^c_{\mathbb{C}}\left(\Omega^1_{X/\mathbb{C},x}, \mathbb{C}\right)\right\}.$$

If $f \in \underline{\underline{\mathbb{C}}}(X)$ represents an element in $\underline{\underline{T}}Z^1(X)_x$, then the corresponding tangent vector has as image in the RHS of (B.7) the map

$$\omega \to \mathrm{Res}_Y(f\omega), \quad \omega \in \Omega^2_{X/\mathbb{C},x}.$$

Being in the kernel on the right-hand side represents compatibility conditions of the form $\sum \mathrm{Res} = 0$ that arise when the local geometry of a family of divisors changes in the limit as $t \to 0$.

AFTERWORD (NOT ESSENTIAL FOR WHAT FOLLOWS)

We will discuss the behavior of the residue map under two sample changes of retraction and verify that the maps in (i) and (ii) below transform correctly.

Let $Y \subset X$ be an irreducible curve. Upon choice of a retraction of a Zariski neighborhood of Y, we have defined a mapping

$$\underline{\underline{T}}Z^1(X) \to \mathrm{Hom}_{\mathcal{O}_Y}\left(\Omega^2_{X/\mathbb{C},Y}, \Omega^1_{\mathbb{C}(Y)/\mathbb{C}}\right).$$

In simplest terms this map goes as follows: The map

(i) $$\underline{\underline{\mathbb{C}}}(X) \to \underline{\underline{T}}Z^1(X)$$

is given in the provisional definition of $\underline{\underline{T}}Z^1(X)$, and composite of this and the above map

$$\underline{\underline{\mathbb{C}}}(X) \to \mathrm{Hom}_{\mathcal{O}_Y}\left(\Omega^2_{X/\mathbb{C},Y}, \Omega^1_{\mathbb{C}(X)/\mathbb{C}}\right)$$

is given by

$$f(\omega) = \operatorname{Res}_Y(f\omega), \quad \omega \in \Omega^2_{X/\mathbb{C},x}.$$

On the other hand, an arc Z_t in $Z^1(X)$ with $Z_0 = kY$ may be thought of as given by a divisor

$$\mathcal{Z} \in X \times B$$

with $\mathcal{Z} \cdot (X \times \{t\}) = Z_t$. We may pull ω back to $X \times B$, restrict it to $\mathcal{Z}$, and write

$$\omega \mid_{\mathcal{Z}} = \varphi \wedge dt,$$

where $\varphi \in \Omega^1_{\mathbb{C}(Y)/\mathbb{C}}$.[5] This gives another map

(ii) $$\underline{\underline{T}} Z^1(X) \to \operatorname{Hom}_{\mathcal{O}_Y}\left(\Omega^2_{X/\mathbb{C},Y}, \Omega^1_{\mathbb{C}(Y)/\mathbb{C}}\right),$$

and these maps agree after suitable identifications are made. However, it is instructive and a good consistency check to verify that they transform the same way when we change retractions.

Assume that we have local uniformizing parameters ξ, η such that Y is given locally by $\xi = 0$. Since the case of $f \in \mathbb{C}(X)$ with a first order pole along Y corresponds to a Poincaré residue where no retraction is necessary, we consider

$$f = \frac{A_2(\eta)}{\xi^2} + \frac{A_1(\eta)}{\xi} + \cdots$$

(this is the case $k = 2$ above). We take the family Y_t to be given by

$$\begin{cases} \xi(t) = a_1(\eta)t^{1/2} + a_2(\eta)t, \\ \eta(t) = \eta. \end{cases}$$

For $\omega = (B_0(\eta) + B_1(\eta)\xi + \cdots)d\xi \wedge d\eta$

(iii) $$\operatorname{Res} f\omega = (A_2 B_1 + A_1 B_0)d\eta.$$

On the other hand, the mapping (ii) above is given by substituting $\xi(t), \eta(t)$ in ω and taking the coefficient of dt. Denoting by $\dot{Y}$ the image in $\operatorname{Hom}_{\mathcal{O}_Y}(\Omega^2_{X/\mathbb{C},Y}, \Omega^1_{\mathbb{C}(Y)/\mathbb{C}})$ of the first-order variation of Y, this gives along $\xi = 0$

(iv) $$\dot{Y}(\omega) = (2B_0 a_2 + B_1 a_1^2)d\eta.$$

Comparing (iii) and (iv) we find, for the given retraction,

(v) $$\begin{cases} A_2 = a_1^2, \\ A_1 = 2a_2. \end{cases}$$

We will consider two cases of a change of retraction.

[5] This procedure may break down when dt vanishes along a component of the support of $\mathcal{Z} \cdot (X \times \{t_0\})$. When this happens a perturbation argument may be used to calculate the map (ii). This phenomenon will be further discussed and illustrated in section 8.3 below.

Case 1:

$$\begin{aligned}\xi &= \xi^{\#}, \\ \eta &= \eta^{\#} + e_1(\eta)\xi + \cdots .\end{aligned}$$

Computation gives

$$\text{(vi)} \qquad \begin{aligned} A_2^{\#} &= A_2, \\ A_1^{\#} &= A_1 + e_1 A_2', \end{aligned}$$

where both sides are considered as functions of $\eta^{\#}$ and $'$ denotes the derivative with respect to η, and also

$$\omega = \left(B_0 + \left(B_0' + B_1 + e_1' B_0\right)\xi^{\#}\right) d\xi^{\#} \wedge d\eta^{\#}.$$

Then

$$\operatorname{Res}_Y f^{\#}\omega^{\#} = \left[A_2\left(B_1 + B_0' - e_1' B_0\right) + \left(e_1 A_2' + A_1\right) B_0\right] d\eta^{\#}.$$

Another computation gives

$$\text{(viii)} \qquad \begin{aligned} \xi^{\#}(t) &= a_1 t^{1/2} + (a_1 a_1' e_1 + a_2)t, \\ a_1^{\#} &= a_1, \\ a_2^{\#} &= a_2 + a_1 a_1' e_1. \end{aligned}$$

Comparing (v)–(vii) gives $A_2^{\#} = (a_1^{\#})^2$ and

$$\begin{aligned} A_1^{\#} &= 2a_2^{\#} = 2\left(a_2 + a_1 a_1' e_1\right), \\ A_1^{\#} &= A_1 + e_1 A_2' = 2a_2 + 2e_1 a_1 a_1' \end{aligned}$$

and we have agreement.

Case 2: For this we take

$$\begin{aligned}\eta &= \eta^{\#}, \\ \xi &= c_1\xi^{\#} + c_2\xi^{\#2} + \cdots, \qquad c_i = c_i(\eta^{\#}) \text{ and } c_1 \neq 0.\end{aligned}$$

Then computation gives

$$\begin{cases} A_2^{\#} = \frac{A_2}{c_1^2}, \\ A_1^{\#} = \frac{A_1}{c_1} - 2\frac{c_2 A_2}{c_1^2}, \end{cases}$$

and

$$\begin{cases} a_1^{\#} = \frac{a_1}{c_1}, \\ a_2^{\#} = \frac{a_2}{c_1} - \frac{c_2 a_1^2}{c_1^3}. \end{cases}$$

As before, one verifies directly that we have agreement.

Conclusion: *The use of a choice of retraction to give the identification (i), (ii) above transforms in such a way as to give a well-defined map*

$$\underline{\underline{T}}Z^1(X) \to \bigoplus_Y \underline{\underline{\mathrm{Hom}}}^{o}_{\mathcal{O}_Y}\left(\Omega^2_{X/\mathbb{C},Y}, \Omega^1_{\mathbb{C}(Y)/\mathbb{C}}\right).$$

Finally, we want to observe that the above discussion implies a global *existence* result concerning the mapping (B.7). Namely, we have that

(B.8) *The map*

$$\underline{\underline{T}}Z^1(X) \to \ker\Big\{ \bigoplus_{Y \text{irred}} \underline{\underline{\mathrm{Hom}}}_{\mathcal{O}_Y}\left(\Omega^2_{X/\mathbb{C},Y}, \Omega^1_{\mathbb{C}(Y)/\mathbb{C}}\right) \to \bigoplus_{x\in X} \underline{\underline{\mathrm{Hom}}}^c\left(\Omega^2_{X/\mathbb{C},x_i}, \mathbb{C}\right)\Big\}$$

is surjective.

Again, we know this to be the case *if* we assume results from general duality theory. The geometric existence result is the surjectivity locally in the Zariski topology of the map

$$\{\text{arcs in } Z^1(X)\} \to \mathcal{PP}_X.$$

So the issue is not so much as to whether or not (B.8) is true as to understand it geometrically, which we now shall discuss. Concretely what is needed is to produce homomorphisms

$$\text{(B.9)} \qquad \varphi_i : \Omega^2_{X,\mathbb{C},x} \to \Omega^1_{\mathbb{C}(Y_i)/\mathbb{C}}, \qquad i = 1, \cdots, n,$$

where the Y_i are algebraic curves passing through x and where the images

$$\varphi_i\left(\Omega^2_{X/\mathbb{C},x_i}\right)$$

have arbitrary poles subject only to the constraint in (B.8). Since poles can only be produced by families of curves acquiring "additional singularities" at x,[6] the issue is one of having "sufficiently many degenerations."

Let $\xi, \eta \in \mathbb{C}(X)$ give local uniformizing parameters at x. We consider the family

$$\xi^k \eta^l = t$$

and the resulting homomorphisms (B.9) on the components $\xi = 0$ and $\eta = 0$ of the limit 1-cycle. To compute these maps we give the above family by

$$\mathcal{Z} \subset X \times B$$

and for $\omega \in \Omega^2_{X/\mathbb{C},x}$ we map

$$\omega \longrightarrow \frac{\partial}{\partial t} \rfloor \left(\pi_X^* \omega \mid_{\mathcal{Z}}\right)$$

restricted to $t = 0$. Up to irrelevant constants this gives

$$\xi^a \eta^b d\xi \wedge d\eta \mapsto \begin{cases} \eta^{b-l} d\eta \text{ on } \xi = 0 \text{ if } a = k-1, 0 \text{ otherwise} \\ \xi^{a-k} d\xi \text{ on } \eta = 0 \text{ if } b = l-1, 0 \text{ otherwise.} \end{cases}$$

It follows that we can get arbitrary poles on $\xi = 0$ and zero on $\eta = 0$, and this implies the surjectivity of (B.8).

[6]We assume that the retraction is smooth in a neighborhood of x, as singularities produced by singularities of the retractions are non-intrinsic and therefore should not be counted.

8.3 DEFINITIONS OF $\underline{\underline{T}}Z_1^1(X)$ FOR X A CURVE AND A SURFACE

To understand geometrically the space of divisors on an algebraic curve, one introduces the relation of linear equivalence $\sim$ and then the quotient (Divisors modulo $\sim$) is the Jacobian variety of the curve. Infinitesimally, letting "T" denote "passing to tangents," the subspace $T(\sim) \subset T$ (Divisors) is defined by the equations

$$\omega = 0, \tag{8.9}$$

where ω varies over the regular 1-forms on the curve.

To extend this picture to 0-cycles on a surface X, we have defined $TZ^2(X)$ and now must define the subspace

$$TZ^2_{\text{rat}}(X) \subset TZ^2(X)$$

of tangents to rational equivalences. By a rational equivalence on the surface X, we mean the following: Setting

$$Z_1^1(X) = \bigoplus_{\begin{cases} Y \text{ codim } 1 \\ Y \text{ irred} \end{cases}} \mathbb{C}(Y)^*,$$

there is a natural map

$$Z_1^1(X) \xrightarrow{\text{div}} Z^2(X) \tag{8.10}$$

given by

$$\sum_\nu (Y_\nu, f_\nu) \to \sum_\nu \text{div } f_\nu,$$

where Y_ν is an irreducible curve on X and $f_\nu \in \mathbb{C}(Y_\nu)^*$ is a rational function on Y_ν. The image of the map (8.10) will be denoted by $Z^2_{\text{rat}}(X)$, the subgroup of 0-cycles that are rationally equivalent to zero. We will define the tangent space $TZ_1^1(X)$ and compute the differential

$$TZ_1^1(X) \to TZ^2(X) \tag{8.11}$$

of the map (8.10). In section 8.4 below, we will give the extension of (8.9) to 0-cycles on surfaces thereby giving Hodge-theoretically the equations that define

$$TZ^2_{\text{rat}}(X) = \{\text{image of (8.11)}\}.$$

This latter point is important, because in contrast to the case of divisors on curves the map (8.10) is not injective. Its kernel, which we may think of as the *irrelevant rational equivalences*, enters into the definition of the higher Chow groups $CH^2(X, 1)$ and is the factor most likely responsible for the difficulty in proving some of the conjectures about rational equivalence in higher codimension, for the usual reason that it is hard to prove existence without a way to control the lack of uniqueness.

The main result of this book will follow from the global consequences of the Cousin flasque resolution of $\Omega^1_{X/\mathbb{Q}}$ together with the following theorem, whose proof will be given below following the discussion of a number of examples that will show what is going on.

(8.12) **Theorem:** *With the formal definition (8.14) below of the tangent sheaf $\underline{\underline{T}} Z_1^1(X)$, the maps that assign to an arc its tangent*

$$\left\{ \text{arcs in } \bigoplus_{\left\{\substack{Y \text{codim } 1 \\ Y \text{ irred}}\right.} \underline{\underline{\mathbb{C}}}(Y)^* \right\} \to \underline{\underline{T}} Z_1^1(X)$$

is surjective.

We emphasize that this is a geometric existence result, albeit one that is local in the Zariski topology and at the infinitesimal level.

We begin with the following

Definition: *For X a smooth algebraic variety of dimension n, we denote by $V^p(X)$ the set of irreducible codimension-p subvarieties of X and define the sheaf*

$$\underline{\underline{Z}}_q^p(X) = \bigoplus_{Y \in V^p(X)} \underline{\underline{K}}_q(\mathbb{C}(Y)).$$

Here, $\underline{\underline{K}}_q(\mathbb{C}(Y))$ is the qth Milnor K-group of the field $\mathbb{C}(Y)$ considered as a constant sheaf on Y and extended to zero outside Y (the proper notation would be $(j_Y)_* \underline{\underline{K}}_q(\mathbb{C}(Y))$). We also set

$$Z_q^p(X) = H^0\left(\underline{\underline{Z}}_q^p(X)\right).$$

As special cases we have

(i)
$$Z_0^p(X) = Z^p(X),$$

(ii)
$$\underline{\underline{Z}}_1^1(X) = \bigoplus_{x \in X} \underline{\underline{\mathbb{C}}}_x^* \qquad (\dim X = 1),$$

and for any n

(iii)
$$\underline{\underline{Z}}_1^1(X) = \bigoplus_{\left\{\substack{\text{codim } Y=1 \\ Y \text{ irred}}\right.} \underline{\underline{\mathbb{C}}}(Y)^*.$$

Note that

$$\begin{aligned} Z_1^1(X) &= H^0(\underline{\underline{Z}}_1^1(X)) \\ &= \{\text{rational equivalences on } X \text{ of codimension 2 cycles}\}. \end{aligned}$$

Finally, we will denote by

$$\underline{\underline{T}} Z_q^p(X) =: T \underline{\underline{Z}}_q^p(X)$$

the tangent sheaf—to be defined below when $p = q = 1$. By (i) above we have already defined $\underline{\underline{T}} Z_0^1(X)$ for any n and $\underline{\underline{T}} Z_0^2$ for $n = 2$.

We have also given in section 6.2 a geometric definition of $\underline{\underline{T}} Z_1^1(X)$ when X is a smooth algebraic curve. There we explained how a set of geometric axioms for first order equivalence of arcs in $Z_1^1(X)$ led to the defining relations for absolute Kähler differentials, thereby introducing $\Omega^1_{X/\mathbb{Q}} = \Omega^1_{\mathcal{O}(X)/\mathbb{Q}}$ into the picture (cf. (6.9)). The definition (6.8) of $\underline{\underline{T}} Z_1^1(X)$ may be expressed by (see below)

$$\underline{\underline{T}} Z_1^1(X) = \bigoplus_{x \in X} H_x^1(\Omega^1_{X/\mathbb{Q}}), \tag{8.13}$$

and for a smooth algebraic surface X we give the analogue

Definition: *The tangent sheaf* $\underline{\underline{T}} Z_1^1(X)$ *is defined by*

$$\underline{\underline{T}} Z_1^1(X) = \bigoplus_{\left\{\substack{Y \text{ codim } 1 \\ Y \text{ irred}}\right.} H_y^1\left(\Omega^1_{X/\mathbb{Q}}\right), \tag{8.14}$$

where y is the generic point of Y.

To justify this definition we check that (8.13) coincides with (6.8) in the curve case. The stalk at x of (8.13) is

$$\lim_k \mathcal{E}xt^1_{\mathcal{O}_X}\left(\mathcal{O}_X/\mathfrak{m}_x^k, \Omega^1_{X/\mathbb{Q}}\right).$$

If ξ is a local uniformizing parameter centered at x, then an element of this group is given by

$$\begin{cases} F_1 \xrightarrow{\xi^k} F_0 \longrightarrow \mathcal{O}_X/\mathfrak{m}_x^k, & F_1, F_0 \cong \mathcal{O}_{X,x}, \\ F^1 \xrightarrow{\psi} \Omega^1_{X/\mathbb{Q},x}. \end{cases}$$

This maps to $\mathrm{Hom}^o(\Omega^1_{X/\mathbb{Q},x}, \Omega^1_{\mathbb{C}/\mathbb{Q}})$ by

$$\varphi \mapsto \mathrm{Res}_x\left(\frac{\psi \wedge \varphi}{\xi^k}\right).$$

Here the residue of a form in $\mathbb{C}(X) \otimes \Omega^2_{X/\mathbb{Q},x}$ is defined as in section 6.3. If $\varphi = f\alpha$, where $f \in \mathcal{O}_{X,x}$, then clearly

$$f\alpha \mapsto \mathrm{Res}_x\left(\frac{f\psi}{\xi^k}\right)\alpha = \varphi_0(f)\alpha,$$

where $\varphi_0(f)$ is defined by the term in the parentheses, and so we have a well-defined map

$$\lim_k \mathcal{E}xt^1_{\mathcal{O}_X}\left(\mathcal{O}_X/\mathfrak{m}_x^k, \Omega^1_{X/\mathbb{Q}}\right) \to \mathrm{Hom}^o\left(\Omega^1_{X/\mathbb{Q},x}, \Omega^1_{\mathbb{C}/\mathbb{Q}}\right). \tag{8.15}$$

We may see that (8.15) is an isomorphism as follows: Using the exact sequence

$$0 \to \mathcal{O}_{X,x} \otimes \Omega^1_{\mathbb{C}/\mathbb{Q}} \xrightarrow{i} \Omega^1_{X/\mathbb{Q},x} \to \Omega^1_{X/\mathbb{C},x} \to 0$$

and the fact that, by definition, the inclusion i induces the map

$$\mathrm{Hom}^o\left(\Omega^1_{X/\mathbb{Q},x}, \Omega^1_{\mathbb{C}/\mathbb{Q}}\right) \to \mathrm{Hom}^c\left(\mathcal{O}_{X,x}, \mathcal{O}_{X,x}\right) \otimes \mathrm{Id}_{\Omega^1_{\mathbb{C}/\mathbb{Q}}},$$

we are reduced to showing that

$$\lim_k \mathcal{E}xt^1_{\mathcal{O}_X}\left(\mathcal{O}_X/\mathfrak{m}_x^k, \mathcal{O}_X\right) \otimes \Omega^1_{\mathbb{C}/\mathbb{Q}} \cong \mathrm{Hom}^c\left(\Omega^1_{X/\mathbb{C},x}, \mathbb{C}\right) \otimes \Omega^1_{\mathbb{C}/\mathbb{Q}}$$

and

$$\lim_k \mathcal{E}xt^1_{\mathcal{O}_X}\left(\Omega^1_{X/\mathbb{C}}/\mathfrak{m}^k_x\Omega^1_{X/\mathbb{C}}, \mathcal{O}_X\right) \cong \operatorname{Hom}^c\left(\mathcal{O}_{X,x}, \mathbb{C}\right),$$

both of which follow from local duality (and of course may be checked directly). Thus for curves the definition (8.13) agrees with the definition (6.8) given earlier.

To justify the definition (8.14) we need to discuss the geometry behind it. One very interesting point is that based on the discussion in section 8.1 above, one might think that the general analogue of

$$\underline{\underline{T}}Z^1_0(X) = \lim_{\left\{\substack{Z \text{ codim } 1 \\ \text{subscheme}}\right.} \mathcal{E}xt^1_{\mathcal{O}_X}\left(\mathcal{O}_Z, \mathcal{O}_X\right) \tag{8.16}$$

would be to define

$$\underline{\underline{T}}Z^1_1(X) = \lim_{\left\{\substack{Z \text{ codim } 1 \\ \text{subscheme}}\right.} \mathcal{E}xt^1_{\mathcal{O}_X}\left(\mathcal{O}_Z, \Omega^1_{X/\mathbb{Q}}\right). \tag{8.17}$$

But for interesting geometric reasons this is *not* correct. As discussed above in section 8.1 and in Appendix B, definition (8.16) reflects compatibility conditions that arise when a reducible curve smooths to an irreducible one. However, such compatibility conditions do *not* arise for arcs in $Z^1_1(X)$. For example, consider the arc

$$\left(\xi\eta = t, \left.\frac{\xi^2+\eta^2}{\xi^2-\eta^2}\right|_{\xi\eta=t}\right)$$

in $\underset{Y}{\oplus}\mathbb{C}(Y)^*$. We are working locally in the Zariski topology where $\xi, \eta \in \mathcal{O}_{X,x}$ give local uniformizing parameters. The first component is the irreducible curve Y_t (or sum of curves at $t = 0$) given by $\xi\eta = t$, and the second component is the designated function in $\mathbb{C}(Y_t)^*$. At $t = 0$ we get the sum

$$(\xi = 0, -1) + (\eta = 0, +1)$$

in $\underset{Y}{\oplus}\mathbb{C}(Y)^*$. At the point of intersection the limit functions do not agree, and hence there are no compatibility conditions. Below, this point will be further discussed and illustrated by differential form calculations. Remark that the right-hand side of (8.17) is equal to

$$T\left(\ker\left(\underline{\underline{Z}}^1_1(X) \xrightarrow{\text{div}} \underline{\underline{Z}}^2(X)\right)\right).$$

Thus

$$\lim_{\left\{\substack{Z \text{ codim} \\ Z \text{ subscheme}}\right.} \mathcal{E}xt^1_{\mathcal{O}_X}\left(\mathcal{O}_Z, \Omega^1_{X/\mathbb{Q}}\right)$$

turns up naturally when one is trying to understand geometrically the tangent spaces to the higher Chow groups.

Before turning to the differential form calculations we shall amplify the definition (8.14) and give a useful expression for calculating the tangent to an arc in $Z^1_1(X)$ when its equations are known.

By definition of local cohomology, we may alternatively give (8.14) as

$$\underline{\underline{T}}Z_1^1(X) = \bigoplus_{\begin{cases} Y \text{ codim } 1 \\ Y \text{ irred} \end{cases}} \lim_{\begin{cases} U \text{ a Zariski} \\ \text{open with } U\cap Y\neq\emptyset \end{cases}} \lim_{k\to\infty} \mathcal{E}xt^1_{\mathcal{O}_U}\left(\mathcal{O}_U/\mathcal{I}_Y^k, \Omega^1_{U/\mathbb{Q}}\right). \tag{8.18}$$

An element in the stalk at $x \in X$ of the term corresponding to an irreducible curve Y passing through x and with $f \in \mathcal{O}_{X,x}$ generating $\mathcal{I}_{Y,x}$ may be described by the data

$$\begin{cases} F_1 \xrightarrow{f^k} F_0 \to \mathcal{O}_X/\mathcal{I}_Y^k, & F_0, F_1 \cong \mathcal{O}_X, \\ F_1 \xrightarrow{g} \Omega^1_{X/\mathbb{Q}}, \end{cases} \tag{8.19}$$

where g does not have Y as a component of its polar locus — this is entirely analogous to the description of $H^1_y(\mathcal{O}_Y)$ used in $\underline{\underline{T}}Z^1(X)$.

We now turn to the description of the tangent map. For this we give an arc in $Z_1^1(X)$ by

$$\left(\operatorname{div}(f + t\dot{f}), (g + t\dot{g})\Big|_{\operatorname{div}(f+t\dot{f})}\right).$$

The first factor describes an arc in the space of 1-cycles, and the second factor gives rational functions on the irreducible components of div $(f + t\dot{f})$. Suppose now that $f \in \mathcal{O}_{X,x}$ with div $f = kY$, where Y is an irreducible curve. Then an element of $\mathcal{E}xt^1_{\mathcal{O}_X}\left(\mathcal{O}_X/\mathcal{I}_Y^k, \Omega^1_{X/\mathbb{Q}}\right)_x$ is given by the data (8.19)

$$\begin{cases} F_1 \xrightarrow{f} F_0 \longrightarrow \mathcal{O}_X/\mathcal{I}_Y^k, \\ F_1 \xrightarrow{\frac{\dot{g}}{g}df - \dot{f}\frac{dg}{g}} \Omega^1_{X/\mathbb{Q}}, \end{cases} \tag{8.20}$$

where $\frac{\dot{g}}{g}df - \dot{f}\frac{dg}{g} \in \Omega^1_{\mathbb{C}(X)/\mathbb{Q}}$ has polar locus not containing Y. This description of the tangent map will agree with the one using differential forms when the appropriate identifications are made.

We now explain, on the sheaf level and with the definition (8.14), what is necessary to establish the surjectivity of the tangent map

$$\{\text{arcs in } Z_1^1(X)\} \to \underline{\underline{T}}Z_1^1(X). \tag{8.21}$$

Namely, let $x \in X$ and set

$$\Omega^1_{X/\mathbb{Q}}\Big|_x = \Omega^1_{X/\mathbb{Q},x}/\mathfrak{m}_x\Omega^1_{X/\mathbb{Q},x}.$$

Let Y be an irreducible curve with $x \in Y$. Then in (8.20) the differential form $(\dot{g}df - \dot{f}dg)/g$ gives by restriction along Y and evaluation at x an element

$$\frac{1}{g}\left(\dot{g}df - \dot{f}dg\right)\Big|_x \in \mathbb{C}(Y) \otimes \Omega^1_{X/\mathbb{Q}}\Big|_x.$$

After an illustrative discussion we shall give a proof of the

(8.22) **Theorem:** *The mapping*

$$\bigoplus_Y \left\{(\dot{f}, \dot{g}) \mapsto \frac{1}{g}(\dot{g}df - \dot{f}dg)\Big|_x\right\} \in \bigoplus_Y \left\{\mathbb{C}(Y) \otimes \Omega^1_{X/\mathbb{Q}}\Big|_x\right\}$$

is surjective.

Here there is a crucial geometric subtlety:

> *In (8.22) one must use the direct sum over all irreducible curves Y to have surjectivity; for a fixed Y the map fails to be surjective.*

Concretely, for smooth Y if we only deform Y to nearly smooth curves we will end up only in part of $\mathbb{C}(Y) \otimes \Omega^1_{Y/\mathbb{Q},x}\big|_x$; to produce forms whose restriction to Y have poles at x we must deform Y into a reducible curve (see below for the proof). When this happens the notation

$$(\dot{f}, \dot{g}) \to \bigoplus_Y \left\{ \frac{1}{g}(\dot{g}df - \dot{f}dg)\Big|_x \right\}$$

needs explanation, which also will be given in equation (vii) below.

Before turning to the formal proof of (8.22), we will continue in our discussion of the geometry behind the definition (8.14).

As in section 8.1 and 8.2 above where we discussed $\underline{\underline{T}}Z^1(X)$, using differential forms gives a good way of understanding this. Write the individual terms in (8.14) as $H^1_y(\Omega^1_{X/\mathbb{Q}})$ and, after choosing for each Y a retraction as explained in Appendix A, use local duality to re-express these terms as

$$\operatorname{Hom}^o_{\mathcal{O}_Y}\left(\Omega^2_{X/\mathbb{Q},Y}/\Omega^2_{\mathbb{C}/\mathbb{Q}}, \Omega^1_{\mathbb{C}(Y)/\mathbb{C}} \otimes \Omega^1_{\mathbb{C}/\mathbb{Q}}\right). \tag{8.23}$$

Here, $\operatorname{Hom}^o_{\mathcal{O}_Y}(\cdot,\cdot)$ has the following meaning: There is an inclusion

$$\Omega^1_{X/\mathbb{C},Y} \otimes \Omega^1_{\mathbb{C}/\mathbb{Q}} \hookrightarrow \Omega^2_{X/\mathbb{Q},Y}/\Omega^2_{\mathbb{C}/\mathbb{Q}}$$

arising from the exact sequence

$$0 \to \Omega^1_{\mathbb{C}/\mathbb{Q}} \otimes \mathcal{O}_X \to \Omega^1_{X/\mathbb{Q}} \to \Omega^1_{X/\mathbb{C}} \to 0.$$

Then

$$\varphi : \Omega^2_{X/\mathbb{Q},Y}/\Omega^2_{\mathbb{C},\mathbb{Q}} \to \Omega^1_{\mathbb{C}(Y)/\mathbb{C}} \otimes \Omega^1_{\mathbb{C}/\mathbb{Q}}$$

is in $\operatorname{Hom}^o_{\mathcal{O}_Y}(\cdot,\cdot)$ if it is $\mathcal{O}_Y$-linear and if for $\omega \in \Omega^1_{X/\mathbb{C},Y}$ and $\alpha \in \Omega^1_{\mathbb{C}/\mathbb{Q}}$

$$\varphi(\omega \otimes \alpha) = \varphi^0(\omega) \otimes \alpha$$

for some $\mathcal{O}_Y$-linear map

$$\varphi^0 : \Omega^1_{X/\mathbb{C},Y} \to \Omega^1_{\mathbb{C}(Y)/\mathbb{C}}. \tag{8.24}$$

Using (8.23) in (8.14) we thus have, upon choice of retractions,

$$\underline{\underline{T}}Z^1_1(X) \cong \bigoplus_Y \underline{\underline{\operatorname{Hom}}}^o_{\mathcal{O}_Y}\left(\Omega^2_{X/\mathbb{Q},Y}/\Omega^2_{\mathbb{C}/\mathbb{Q}}, \Omega^1_{\mathbb{C}(Y)/\mathbb{C}} \otimes \Omega^1_{\mathbb{C}/\mathbb{Q}}\right). \tag{8.25}$$

Although the notation is a bit complicated, this identification contains a lot of geometry. We shall illustrate this through several examples and special cases.

(a) One of the equivalent prescriptions for computing the tangent map

$$\text{arcs in } Z^1_1(X) \to \underline{\underline{T}}Z^1_1(X) \tag{8.26}$$

is the following: Let B denote the parameter curve with parameter $t \in B$ and represent an arc in $Z^1_1(X)$ by a codimension-2 cycle

$$\mathcal{Z} \subset X \times \mathbb{P}^1 \times B$$

where

$$\mathcal{Z} \cdot \left(X \times \mathbb{P}^1 \times \{t\}\right) = \sum_i n_i \left(Y_{i,t}, f_{i,t}\right)$$

with the $Y_{i,t}$ being irreducible curves, $f_{i,t} \in \mathbb{C}(Y_{i,t})^*$, and $(Y_{i,t}, f_{i,t})$ the graph of $f_{i,t}$ in $X \times \mathbb{P}^1$. To compute the image of this curve in the stalk at $x \in X$ of this arc in $Z^1_1(X)$, we let $u \in \mathbb{C}^* \subset \mathbb{P}^1$ be a standard coordinate and $\omega \in \Omega^2_{X/\mathbb{Q},x}$. We then write

$$\omega \wedge \left.\frac{du}{u}\right|_{\mathcal{Z}} = \tau(\omega) \wedge dt$$

and the map (8.26) is given by

$$\omega \to \tau(\omega)\big|_{t=0}. \tag{8.27}$$

Here, the notation $\tau(\omega)\big|_{t=0}$ is understood as follows: If Y_i are the components in X of $\mathcal{Z} \cdot (X \times \mathbb{P}^1 \times \{0\})$, then the restriction of $\tau(\omega)$ to Y_i is in $\Omega^2_{\mathbb{C}(Y_i)/\mathbb{Q}}$. Since Y_i is an algebraic curve, there is for dimension reasons a natural map

$$\Omega^2_{\mathbb{C}(Y_i)/\mathbb{Q}} \to \Omega^1_{\mathbb{C}(Y_i)/\mathbb{C}} \otimes \Omega^1_{\mathbb{C}/\mathbb{Q}}$$

and the image of $\tau(\omega)$ under this map is what is meant by (8.27).

There is one caveat to this prescription: Namely, we may have $dt = 0$ along some of the components of $\mathcal{Z} \cdot (X \times \mathbb{P}^1 \times \{0\})$. When this happens

$$\partial/\partial t \rfloor \left(\omega \wedge \left.\frac{du}{u}\right|_{\mathcal{Z}}\right)$$

will have a pole on the projection to X of that component, so that (8.27) is not defined. What one does in this case is to perturb $\mathcal{Z}$ to a family $\mathcal{Z}_\lambda$ where $\mathcal{Z}_0 = \mathcal{Z}$ and where (8.27) is well-defined for $\lambda \neq 0$, and then take the limit of (8.27) as $\lambda \to 0$ (essentially this is l'Hôpital's rule). This procedure will be illustrated by example below.

Remark that this last phenomenon corresponds (also see below) to the case of taking the residue of a form with a higher order pole along a curve. It is here that the retraction is used and this corresponds to making a perturbation as described above. All of this will be quite clear in the example below.

(b) We may define a map

$$\Omega^1_{\mathbb{C}(X)/\mathbb{Q}} \xrightarrow{dT} \bigoplus_Y \underline{\underline{\mathrm{Hom}}}^{o}_{\mathcal{O}_Y} \left(\Omega^2_{X/\mathbb{Q},Y}/\Omega^2_{\mathbb{C}/\mathbb{Q}}, \Omega^1_{\mathbb{C}(Y)} \otimes \Omega^1_{\mathbb{C}/\mathbb{Q}}\right) \tag{8.28}$$

by sending, for $\psi \in \Omega^1_{\mathbb{C}(X)/\mathbb{Q}}$ and $\omega \in \Omega^2_{X/\mathbb{Q},Y}$,

$$\psi \otimes \omega \longrightarrow \mathrm{Res}_Y(\psi \wedge \omega).$$

Here the residue is defined as in section 6.3 extended one dimension up and using the retraction as in section 8.1 above.

(8.29) **Proposition:** *Using the identification (8.25) and (cf. the Appendix to section 6.3)*

$$\underline{\underline{T}}K_2(\mathbb{C}(X)) \cong \Omega^1_{\mathbb{C}(X)/\mathbb{Q}}$$

the map (8.28) may be identified with the differential of the tame symbol.

We have in section 6.3 given a proof of this one dimension down. We shall give a different argument based upon (8.20) and illustrate it in an example that exhibits phenomena not encountered in the curve case. From this the result should be clear.

An arc in $K_2(\mathbb{C}(X))$ is given by $\{f + t\dot{f}, g + t\dot{g}\}$, where $f, g \in \mathbb{C}(X)^*$, $\dot{f}, \dot{g} \in \mathbb{C}(X)$, and where we assume that the divisors of $f + t\dot{f}$ and $g + t\dot{g}$ have no curve components in common. Then the tame symbol is given by

$$\{f + t\dot{f}, g + t\dot{g}\} \to g + t\dot{g}\Big|_{\mathrm{div}(f+t\dot{f})} - f + t\dot{f}\Big|_{\mathrm{div}(g+t\dot{g})}.$$

From (6.10), the tangent to this arc in

$$\bigoplus_{\begin{cases} Y \text{ codim } 1 \\ Y \text{ irred} \end{cases}} H^1_y\left(\Omega^1_{X/\mathbb{Q}}\right)$$

is the sum of

$$\begin{cases} F_1 \xrightarrow{f} F_0 \to \mathcal{O}_X/(f) & F_1, F_0 \cong \mathcal{O}_X, \\ F_1 \xrightarrow{\frac{\dot{g}}{g}df - \dot{f}\frac{dg}{g}} \Omega^1_{X/\mathbb{Q}}, & \end{cases}$$

and

$$\begin{cases} F_1' \xrightarrow{g} F_0' \to \mathcal{O}_X/(g) & F_1', F_0' \cong \mathcal{O}_X, \\ F_1' \xrightarrow{-\frac{\dot{f}}{f}dg + \dot{g}\frac{df}{f}} \Omega^1_{X/\mathbb{Q}}. & \end{cases}$$

Here, to be precise we should work in a Zariski neighborhood of a point $x \in X$, factor f, g into irreducible factors in $\mathcal{O}_{X,x}$, and by the methods used in the Appendix to section 6.3 write the symbol $\{f + t\dot{f}, g + t\dot{g}\}$ as a product of symbols with constant t-terms in $\mathcal{O}_{X,x}$.

On the other hand, from the discussion in that Appendix the tangent to the arc $\{f + t\dot{f}, g + t\dot{g}\}$ in $K_2(\mathbb{C}(X))$ is

$$\frac{\dot{g}}{g}\frac{df}{f} - \frac{\dot{f}}{f}\frac{dg}{g} \in \Omega^1_{\mathbb{C}(X)/\mathbb{Q}}.$$

Under the map

$$\underline{\underline{\Omega}}^1_{\mathbb{C}(X)/\mathbb{Q}} \to \bigoplus_{\begin{cases} Y \text{ codim } 1 \\ Y \text{ irred} \end{cases}} \underline{\underline{H}}^1_y\left(\Omega^1_{X/\mathbb{Q}}\right)$$

this maps to the same element as given above.

Before turning to the example, it is useful to compute in this notation the differential of

$$\bigoplus_{\begin{cases} Y \text{ codim } 1 \\ Y \text{ irred} \end{cases}} \underline{\underline{\mathbb{C}}}(Y)^* \to \bigoplus_{x \in X} \underline{\underline{\mathbb{Z}}}_x,$$

which maps the arc $g+t\dot{g}\big|_{\mathrm{div}(f+t\dot{f})}$ in the first term to the arc $\mathrm{Var}(g+t\dot{g},\ f+t\dot{f})$ in the second. Again, localizing at $x \in X$ and assuming that $f, g \in \mathcal{O}_{X,x}$ are relatively prime, the tangent to the first arc in $\bigoplus_Y H^1_y\left(\Omega^1_{X/\mathbb{Q}}\right)_x$ is

$$\begin{cases} F_1 \xrightarrow{f} F_0 \to \mathcal{O}_X/(f), & F_0, F_1 \cong \mathcal{O}_X, \\ F_1 \xrightarrow{\frac{\dot{g}}{g}df - \frac{\dot{f}dg}{g}} \Omega^1_{X/\mathbb{Q}}. \end{cases}$$

Under the natural map

$$\bigoplus_Y \underline{\underline{H}}^1_y\left(\Omega^1_{X/\mathbb{Q}}\right) \to \bigoplus_{x\in X} \underline{\underline{H}}^2_x\left(\Omega^1_{X/\mathbb{Q}}\right)$$

in the Cousin flasque resolution, this maps to

$$\begin{cases} E_2 \xrightarrow{\binom{g}{-f}} E_1 \xrightarrow{(f\ g)} E_0 \to \mathcal{O}_X/(f,g) & E_2, E_0 \cong \mathcal{O}_X,\ E_1 \cong \mathcal{O}^2_X, \\ E_2 \xrightarrow{\dot{g}df - \dot{f}dg} \Omega^1_{X/\mathbb{Q}}. \end{cases}$$

On the other hand, the family

$$z_t = \mathrm{Var}(f+t\dot{f},\ g+t\dot{g})$$

has a minimal free resolution

$$E'_2 \xrightarrow{\binom{g+t\dot{g}}{-f-t\dot{f}}} E'_1 \xrightarrow{(f+t\dot{f},g+t\dot{g})} E'_0 \to \mathcal{O}_{z_t} \qquad E'_2, E'_0 \cong \mathcal{O}_X,\ E'_1 \cong \mathcal{O}^2_X.$$

By the prescription in chapter 7, its tangent is the dt coefficient of

$$\frac{1}{2}\left(d(f+t\dot{f}), d(g+t\dot{f})\right)\begin{pmatrix} d(g+t\dot{g}) \\ -d(f+t\dot{f}) \end{pmatrix},$$

which is

$$E'_2 \xrightarrow{\dot{g}df - \dot{f}dy} \Omega^1_{X/\mathbb{Q}}$$

in agreement with the above.

Summary: These calculations show that the differentials of the maps in the Bloch-Gersten-Quillen sequence

$$\underline{\underline{K}}_2(\mathbb{C}(X)) \xrightarrow{T} \bigoplus_Y \underline{\underline{\mathbb{C}}}(Y)^* \to \bigoplus_{x\in X} \underline{\underline{\mathbb{Z}}}_x$$

are the maps in the Cousin flasque sequence

$$\underline{\underline{\Omega}}^1_{\mathbb{C}(X)/\mathbb{Q}} \to \bigoplus_Y \underline{\underline{H}}^1_y\left(\Omega^1_{X/\mathbb{Q}}\right) \to \bigoplus_{x\in X} \underline{\underline{H}}^2_x\left(\Omega^1_{X/\mathbb{Q}}\right).$$

Proposition: *The tangent sequence to the Bloch-Gersten-Quillen sequence*

$$0 \to K_2(\mathcal{O}_X) \to \underline{\underline{K}}_2(\mathbb{C}(X)) \to \bigoplus_Y \underline{\underline{\mathbb{C}}}(Y)^* \to \bigoplus_{x\in X} \underline{\underline{\mathbb{Z}}}_x \to 0$$

is the Cousin flasque resolution of $\Omega^1_{X/\mathbb{Q}}$

$$0 \to \Omega^1_{X/\mathbb{Q}} \to \underline{\underline{\Omega}}^1_{\mathbb{C}(X)/\mathbb{Q}} \to \bigoplus_{Y} \underline{\underline{H}}^1_y(\Omega^1_{X/\mathbb{Q}}) \to \bigoplus_{x\in X} \underline{\underline{H}}^2_x(\Omega^1_{X/\mathbb{Q}}) \to 0.$$

Here, we have defined the tangent sheaf to each of the terms in the top sequence, and the proposition states that these may be naturally identified with the terms in the bottom sequence.

We now turn to an example that illustrates phenomena that do not occur in the curve case.

Example: We consider the curve

$$\{\xi^2 + t, \eta\}$$

in $K_2(\mathbb{C}(X))$, where $\xi, \eta \in \mathbb{C}(X)$ give local uniformizing parameters on a Zariski open set in X. From the discussion in the Appendix to section 6.3, the tangent to this curve is

$$\frac{d\eta}{\xi^2\eta} \in \Omega^1_{\mathbb{C}(X)/\mathbb{Q}}.$$

We first compute

$$\mathrm{Res}_{Y_i}\left(\frac{d\eta}{\xi^2\eta} \wedge \omega\right)$$

on the components $Y_0 = \{\eta = 0\}$ and $Y_1 = \{\xi = 0\}$ of the polar locus of the form in parenthesis. We will only get something nonzero in case

$$\omega = f(\xi, \eta) d\xi \wedge d\alpha,$$

where $d\alpha \in \Omega^1_{\mathbb{C}/\mathbb{Q}}$. Then

$$\begin{cases} \mathrm{Res}_{Y_0}\left(\frac{d\eta}{\eta} \wedge \frac{f(\xi,\eta)d\xi\wedge d\alpha}{\xi^2}\right) = \frac{f(\xi,0)}{\xi^2} d\xi \wedge d\alpha, \\ \mathrm{Res}_{Y_1}\left(\frac{d\xi}{\xi^2} \wedge \frac{f(\xi,\eta)d\eta\wedge d\alpha}{\eta}\right) = \frac{f_\xi(0,\eta)}{\eta} d\eta \wedge d\alpha. \end{cases}$$

Taking signs into account the final result is

$$\frac{f(\xi,0)}{\xi^2} d\xi \wedge d\alpha - \frac{f_\xi(0,\eta)}{\eta} d\eta \wedge d\alpha. \tag{8.30}$$

On the other hand, applying the tame symbol T to the above arc in $K_2(\mathbb{C}(X))$ we obtain

$$(\eta = 0, \xi^2 + t) - (\xi^2 + t, \eta). \tag{8.31}$$

To compute the tangent to this arc in $Z^1_1(X)$ we shall use (a) above. For this we let the curve B have parameter t and consider the codimension-2 cycle

$$\mathcal{Z} \subset X \times \mathbb{P}^1 \times B$$

given by (8.31). As before we need only consider ω's of the form $\omega = f(\xi, \eta) d\xi \wedge d\alpha$ where $d\alpha \in \Omega^1_{\mathbb{C}/\mathbb{Q}}$. The prescription in (a) is to take $u \in \mathbb{C}^* \subset \mathbb{P}^1$ as coordinate and then take the restriction to $t = 0$ of

$$\frac{\partial}{\partial t} \lrcorner \left(\omega \wedge \frac{du}{u}\bigg|_{\mathcal{Z}}\right).$$

For the first term in (8.31) this gives

$$f(\xi, 0)\frac{d\xi}{\xi^2} \wedge d\alpha \tag{8.32}$$

on the curve $\eta = 0$. For the second term in (8.31), if we use the relation

$$2\xi d\xi + dt = 0 \tag{8.33}$$

on $\mathcal{Z}$ we have

$$\frac{\partial}{\partial t} \rfloor \left(\omega \wedge \frac{du}{u}\bigg|_{\mathcal{Z}} \right) = -\frac{f(\xi,\eta)}{2\xi}\frac{d\eta}{\eta} \wedge d\alpha,$$

which blows up if we try to set $\xi = 0$. The trouble is that dt vanishes on the component $\mathcal{Z} \cap \{\xi = 0\}$ of $\mathcal{Z} \cdot (X \times \mathbb{P}^1 \times \{0\})$, and so we cannot divide by it.

To resolve this problem, as explained above we perturb our family to one where this does not happen and then take a limit. Thus we consider the curve

$$\{\xi(\xi + \lambda) + t, \eta\}$$

in $K_2(\mathbb{C}(X))$. The first term in the analogue of (8.31) is as before, and the second is now

$$(\xi(\xi + \lambda) + t, \eta).$$

Here we have in place of (8.33)

$$(2\xi + \lambda)d\xi + dt = 0.$$

There are two components of $\mathcal{Z}$ over $t = 0$; they are given by $\xi = 0$ and $\xi + \lambda = 0$. On the first and second respectively of these components

$$\partial/\partial t \rfloor \left(\omega \wedge \frac{du}{u}\bigg|_{\mathcal{Z}} \right) = -\frac{f(0,\eta)}{\lambda}\frac{d\eta}{\eta} \wedge d\alpha,$$

$$\partial/\partial t \rfloor \left(\omega \wedge \frac{du}{u}\bigg|_{\mathcal{Z}} \right) = +\frac{f(-\lambda,\eta)}{\lambda}\frac{d\eta}{\eta} \wedge d\alpha.$$

Adding these and taking the limit as $\lambda \to 0$ gives

$$-f_\xi(0,\eta)\frac{d\eta}{\eta} \wedge d\alpha.$$

Adding this to (8.32) we find agreement with (8.30).

Remark: If we use the relation (8.33), then as noted above

$$\partial/\partial t \rfloor \left(\omega \wedge \frac{du}{u}\bigg|_{\mathcal{Z}} \right) = -\frac{f(\xi,\eta)}{2\xi}\frac{d\eta}{\eta} \wedge d\alpha. \tag{8.34}$$

We write

$$f(\xi,\eta) = f(0,\eta) + \xi f_\xi(0,\eta) + \cdots$$

and substitute in (8.34) and sum up over the two branches of $\xi^2 + t = 0$. Then the constant term in the Taylor expansion cancels out and the linear term gives

$$-f_\xi(0,\eta)\frac{d\eta}{\eta} \wedge d\alpha.$$

Taking the limit as $t \to 0$ we find agreement with the calculation above.

(c) We shall examine the tangents to a number of special kinds of arcs in $Z_1^1(X)$.

The first is an arc (z_t, λ), where z_t is an arc in $Z^1(X)$ and $\lambda \in \mathbb{C}^*$ is a constant. Recalling that for the tangent $\dot{z}$ to z_t we have, upon choices of retractions,

$$\dot{z} \in \bigoplus_{Y \text{ irred}} \operatorname{Hom}^o_{\mathcal{O}_Y}\left(\Omega^2_{X/\mathbb{C},Y}, \Omega^1_{\mathbb{C}(Y)/\mathbb{C}}\right).$$

Denoting by

$$\tau \in \bigoplus_{Y \text{ irred}} \operatorname{Hom}^o_{\mathcal{O}_Y}\left(\Omega^2_{X/\mathbb{Q},Y}/\Omega^2_{\mathbb{C}/\mathbb{Q}}, \Omega^1_{\mathbb{C}(Y)/\mathbb{C}} \otimes \Omega^1_{\mathbb{C}/\mathbb{Q}}\right)$$

the tangent to (z_t, λ) we have

$$\tau = \dot{z} \otimes \frac{d\lambda}{\lambda}, \tag{8.35}$$

where we are using the natural mapping

$$\operatorname{Hom}^o_Y\left(\Omega^2_{X/\mathbb{C},Y}, \Omega^1_{\mathbb{C}(Y)/\mathbb{C}}\right) \to \operatorname{Hom}^o_Y\left(\Omega^2_{X/\mathbb{Q},Y}/\Omega^2_{\mathbb{C}/\mathbb{Q}}, \Omega^1_{\mathbb{C}(Y)/\mathbb{C}}\right)$$

applied to $\dot{z}$. We note that the fact that

$$\dot{z} \in \ker\left\{\bigoplus_{Y} \operatorname{Hom}^o_{\mathcal{O}_Y}\left(\Omega^2_{X/\mathbb{C},Y}, \Omega^1_{\mathbb{C}(Y)/\mathbb{C}}\right) \to \bigoplus_{x \in X} \operatorname{Hom}^o_{\mathbb{C}}\left(\Omega^2_{X/\mathbb{C},x}, \mathbb{C}\right)\right\}$$

is reflected by the obvious fact that under the map

$$Z_1^1(X) \xrightarrow{\text{div}} Z^2(X)$$

the arc (z_t, λ) maps to zero. The geometric reason why the arithmetic properties of λ enter into the definition of the tangent to (z_t, λ) were discussed in section 6.2.

A second special type of arc in $Z_1^1(X)$ is given by (z, f_t), where f_t is an arc in $\mathbb{C}(X)^*$ and where no component of the divisors of the f_t contains a component of the 1-cycle z. Now there is a natural map

$$\operatorname{Hom}_{\mathcal{O}_X}\left(\Omega^1_{X/\mathbb{C},Y}, \Omega^1_{\mathbb{C}(Y)/\mathbb{C}}\right) \xrightarrow{j} \operatorname{Hom}^o_{\mathcal{O}_Y}\left(\Omega^2_{X/\mathbb{Q},Y}/\Omega^2_{\mathbb{C}/\mathbb{Q}}, \Omega^1_{\mathbb{C}(Y)/\mathbb{C}} \otimes \Omega^1_{\mathbb{C}/\mathbb{Q}}\right)$$

given by using the natural map

$$\begin{array}{ccccc} \Omega^2_{X/\mathbb{Q},Y}/\Omega^2_{\mathbb{C}/\mathbb{Q}} & \longrightarrow & \Omega^2_{Y/\mathbb{Q}}/\Omega^2_{\mathbb{C}/\mathbb{Q}} \longrightarrow \Omega^1_{\mathbb{C}(Y)} \otimes \Omega^2_{\mathbb{C}/\mathbb{Q}} \\ \omega & \longrightarrow & \alpha(\omega) \otimes \beta(\omega) \end{array}$$

and sending $\tau \in \operatorname{Hom}_{\mathcal{O}_X}\left(\Omega^1_{X/\mathbb{C},Y}, \Omega^1_{\mathbb{C}(Y)/\mathbb{C}}\right)$ to the map given by

$$j(\tau)(\omega) = \tau(\alpha(\omega)) \otimes \beta(\omega).$$

Then the tangent to (z, f_t) is given by

$$j(\text{restriction to } Y)\frac{\dot{f}}{f},$$

where

$$(\text{restriction to } Y) \in \operatorname{Hom}_{\mathcal{O}_X}\left(\Omega^1_{X/\mathbb{C},Y}, \Omega^1_{\mathbb{C}(Y)/\mathbb{C}}\right)$$

is just the usual restriction mapping. More simply, the map is

$$\omega \longrightarrow \left(\frac{\dot{f}}{f}\alpha(\omega)\right) \otimes \beta(\omega) \in \Omega^1_{\mathbb{C}(Y)/\mathbb{C}} \otimes \Omega^1_{\mathbb{C}/\mathbb{Q}}. \tag{8.36}$$

(d) From the exact sequence

$$0 \to \Omega^1_{X/\mathbb{C}} \otimes \Omega^1_{\mathbb{C}/\mathbb{Q}} \to \Omega^2_{X/\mathbb{Q}}/\Omega^2_{X/\mathbb{C}} \to \Omega^2_{X/\mathbb{C}} \to 0$$

we infer the sequence

$$\begin{aligned} 0 \to \bigoplus_Y \underline{\underline{\mathrm{Hom}}}^{o}_{\mathcal{O}_Y}\left(\Omega^2_{X/\mathbb{C},Y}, \Omega^1_{\mathbb{C}(Y)/\mathbb{C}} \otimes \Omega^1_{\mathbb{C}/\mathbb{Q}}\right) \\ \to \underline{\underline{T}}Z^1_1(X) \xrightarrow{\pi} \bigoplus_Y \underline{\underline{\mathrm{Hom}}}_{\mathcal{O}_Y}\left(\Omega^1_{X/\mathbb{C},Y}, \Omega^1_{\mathbb{C}(Y)/\mathbb{C}}\right) \otimes \mathrm{Id}_{\Omega^1_{\mathbb{C}/\mathbb{Q}}} \to 0. \end{aligned} \tag{8.37}$$

The mapping π may be thought of as follows: *Ignore any arithmetic aspect of an arc in* $Z^1_1(X)$. For example, by (8.36) above tangents to arcs of the form (z, f_t) are sent under π to the map

$$\left(\frac{\dot{f}}{f}\right)(\text{restriction to } Y)\mathrm{Id}_{\Omega^1_{\mathbb{C}/\mathbb{Q}}}.$$

In general we may describe π as follows: Represent an arc in $Z^1_1(X)$ by

$$\mathcal{Z} \subset X \times \mathbb{P}^1 \times B.$$

Then for $\varphi \in \Omega^1_{X/\mathbb{C},x}$ we map

$$\varphi \mapsto \sum_i n_i \left(\partial/\partial t \rfloor \left(\varphi \wedge \frac{du}{u}\Big|_{\mathcal{Z}}\right)\Big|_{Y_i}\right), \tag{8.38}$$

where $\mathcal{Z} \cdot \left(X \times \mathbb{P}^1 \times t_0\right) = \sum_i n_i(Y_i, f_i)$.

The kernel of π in (8.37) is the subspace

$$\begin{array}{c} \bigoplus_Y \underline{\underline{\mathrm{Hom}}}^{c}_{\mathcal{O}_Y}\left(\Omega^2_{X/\mathbb{C},Y}, \Omega^1_{\mathbb{C}(Y)/\mathbb{C}}\right) \otimes \Omega^1_{\mathbb{C}/\mathbb{Q}} \subset \underline{\underline{T}}Z^1_1(X) \\ \uparrow \\ \underline{\underline{T}}Z^1(X) \otimes \Omega^1_{\mathbb{C}/\mathbb{Q}} \end{array} \tag{8.39}$$

Here, the dotted vertical arrow refers to the map

$$\underline{\underline{T}}Z^1(X) \otimes \Omega^1_{\mathbb{C}/\mathbb{Q}} \to \underline{\underline{T}}Z^1_1(X)$$

given as the differential of the map

$$(\text{arc } z_t \text{ in } Z^1(X)) \otimes d\alpha \longrightarrow \text{arc } (z_t, e^{\alpha}) \text{ in } Z^1_1(X)$$

encountered earlier.

We do not know a geometric interpretation of the full subspace (8.39).

(e) In section 8.1 we discussed two properties that $\underline{\underline{T}}Z^1(X)$ might have. Here we shall discuss the analogues for $\underline{\underline{T}}Z^1_1(X)$.

Using local duality and theorem (8.22), the map

$$\{\text{arcs in } Z^1_1(X)\} \to \bigoplus_Y \underline{\underline{\mathrm{Hom}}}^{o}_{\mathcal{O}_Y}\left(\Omega^2_{X/\mathbb{C},Y}/\Omega^2_{\mathbb{C}/\mathbb{Q}}, \Omega^1_{\mathbb{C}(Y)/\mathbb{C}} \otimes \Omega^1_{\mathbb{C}/\mathbb{Q}}\right) \tag{8.40}$$

should be surjective. The proof of (8.22) given below emerged from understanding the following types of examples. As was the case for the surjectivity of the analogous map in section 8.1, the issue is to produce poles. In these illustrative examples we shall let ξ, η be local uniformizing parameters.

(i) For the family given by

$$(\xi + \alpha t^{1/2} = 0, \eta + \beta t^{1/2})$$

we have

$$d\xi \wedge d\eta \mapsto \frac{d\eta}{\eta} \otimes (\alpha d\beta - \beta d\alpha) \in \Omega^1_{\mathbb{C}(Y)/\mathbb{C}} \otimes \Omega^1_{\mathbb{C}/\mathbb{Q}},$$

where Y is given by $\xi = 0$. Here the notation means the following: For each of the two choices of $\sqrt{t}$ we may define a curve $Y_t^{\pm}$ by $\xi \pm \alpha t^{1/2} = 0$, and on $Y_t^{\pm}$ a function given by $\eta \pm \beta t^{1/2}$ using the same value of $\sqrt{t}$. Adding these gives an arc in $Z_1^1(X)$. To compute the tangent to this arc interpreted as an element on the right-hand side in (8.40), we have used the prescription (8.28). Similar considerations apply also to the following examples.

(ii) An example that illustrates the absence of compatibility conditions in $\underline{\underline{T}}Z_1^1(X)$ is given by the arc

$$(\xi\eta - t = 0, \alpha\xi + \beta\eta), \qquad \alpha, \beta \in \mathbb{C}.$$

For its tangent we have

$$d\xi \wedge d\eta \to \begin{cases} -\frac{d\eta}{\eta} \wedge \frac{d\beta}{\beta} \text{ on } \xi = 0, \\ +\frac{d\xi}{\xi} \wedge \frac{d\alpha}{\alpha} \text{ on } \eta = 0. \end{cases}$$

(iii) For the next example we take the arc

$$(\xi^m \eta - t = 0, e^{\alpha}), \qquad \alpha \in \mathbb{C}$$

in $Z_1^1(X)$. A similar argument to that which gave (8.31) gives for

$$\omega = g(\xi)\eta d\xi \wedge d\eta$$

that the tangent to the above arc is

$$\omega \mapsto \begin{cases} \frac{g(\xi)}{\xi^m} d\xi \wedge d\alpha & \text{on } \eta = 0, \\ \left(\frac{1}{m}\right) g^{(m-1)}(0) d\eta \wedge d\alpha & \text{on } \xi = 0. \end{cases}$$

By taking linear combinations of this together with examples (i) and (ii) above we have produced an arbitrary linear combination

$$\sum_{k=m}^{0} \frac{c_k}{\xi^k} d\alpha_k \wedge d\xi, \qquad c_k \in \mathbb{C} \text{ and } d\alpha_k \in \Omega^1_{\mathbb{C}/\mathbb{Q}}.$$

Proof of Theorem (8.22): The proof will proceed in three steps.

Step one: For $x \in Y$ a smooth point we will show that

(i)
$$\{\text{arcs in } Z_1^1(X)\} \to \mathcal{O}_{Y,x} \underset{\mathbb{C}}{\otimes} \left(\Omega^1_{X/\mathbb{Q},x}\big|_x\right)$$

is surjective. (Actually we will prove more—that we can reach first order poles along Y and arbitrary poles in directions normal to Y.)

Step two: For $x \in Y$ a smooth point with local uniformizing parameter ξ, for each n we will produce arcs in $Z_1^1(X)$ whose polar part reaches

(ii)
$$\left(\frac{1}{\xi^n}\right) \otimes \Omega^1_{X/\mathbb{Q},x}\big|_x + \text{lower order terms.}$$

Step three: From the first two steps we have surjectivity in the stalk at x in (8.22) for all those Y for which $x \in Y$ is a smooth point. Using this we will deduce (i) and (ii) above when $x \in Y$ may be a singular point and ξ is a local uniformizing paramater on a point of the normalization $\widetilde{Y} \xrightarrow{\pi} Y$ lying over x.

Step one is easy: If $x \in Y$ is a smooth point then expressions of the form

$$\frac{1}{g}\left(\dot{g}df - \dot{f}dg\right)\Big|_x$$

where $\dot{f}, \dot{g} \in \mathcal{O}_{X,x}$ clearly reach all of $\mathcal{O}_{X,x} \otimes_{\mathbb{C}} \left(\Omega^1_{X/\mathbb{Q},x}\big|_x\right)$. Here taking, for example, $g = e^{\alpha}$ gives

$$\left(-\dot{f}\big|_Y\right) \otimes d\alpha \in \mathcal{O}_{Y,x} \otimes_{\mathbb{C}} \Omega^1_{\mathbb{C},\mathbb{Q}},$$

which reaches the $\Omega^1_{\mathbb{C}/\mathbb{Q}}$-part of $\Omega^1_{X/\mathbb{Q},x}\big|_x$. Actually, as will be seen by explicit computation below, using $\dot{g}df/g$'s we may reach arbitrary poles in the co-normal direction to Y and $\dot{f}dg/g$'s to reach first order poles along Y.

Step two is more interesting and involves one fundamental new point. Thinking of $\underline{\underline{H}}^1_y\left(\Omega^1_{X/\mathbb{Q}}\right)_x$ in terms of $\mathcal{E}xt$'s, for $\varphi \in \Omega^1_{X/\mathbb{Q},x}$ we will let

$$\frac{\varphi}{\mathcal{O}_X \xrightarrow{f} \mathcal{O}_X}$$

denote the element that sends $1 \in \mathcal{O}_{X,x} \cong E_1$ to φ. The denominator means that we impose the equivalence relation

$$\varphi \sim \varphi + f\psi.$$

Working in $\underline{\underline{H}}^1_y(\Omega^1_{X/\mathbb{Q}})_x$ where y is a generic point of Y means that we may allow φ to be a rational 1-form on X whose polar locus does not include Y as a component, and similarly for ψ. Thus the map

$$\frac{\varphi}{\mathcal{O}_X \xrightarrow{f} \mathcal{O}_X} \mapsto \varphi\big|_x \in \mathbb{C}(Y) \otimes_{\mathbb{C}} \Omega^1_{X/\mathbb{Q},x}\big|_x$$

is well-defined. This map means: Write φ as a linear combination of 1-forms in $\Omega^1_{X/\mathbb{Q},x}$ with coefficients in $\mathbb{C}(X)$ not having poles on Y and then restrict the coefficient functions to Y ending up in $\mathbb{C}(Y)$. Since the tensor product is over $\mathbb{C}$ this mapping is well-defined, and essentially we have to show that *it is surjective for those φ's that arise from tangents to arcs in $Z^1_1(X)$.*

We will first work on the image of the map

$$\mathbb{C}(Y) \otimes_{\mathbb{C}} \Omega^1_{X/\mathbb{Q},x}\big|_x \to \mathbb{C}(Y) \otimes_{\mathbb{C}} \Omega^1_{X/\mathbb{C},x}\big|_x .$$

The remaining $\mathbb{C}(Y) \otimes_{\mathbb{C}} \Omega^1_{\mathbb{C}/\mathbb{Q}}$-part may then be treated similarly using considerations as in step one.

Let now $\xi, \eta \in \mathcal{O}_{X,x} \subset \mathbb{C}(X)$ be local uniformizing parameters centered at x and with $\eta = 0$ being a local defining equation for Y. Then for $h \in \mathcal{O}_{X,x}$

$$(\eta - th, g) \to \frac{h\frac{dg}{g}}{\mathcal{O}_X \xrightarrow{\eta} \mathcal{O}_X}.$$

The notation means: The arc $(\eta - th, g)$ in $\bigoplus_Y \mathbb{C}(Y)^*$ maps to the expression on the right in $\underline{\underline{H}}^1_y\left(\Omega^1_{X/\mathbb{C}}\right)_x$. Taking $h = 1$ and $g = \alpha\xi$ we have

$$\text{(iii)} \qquad \alpha\frac{d\xi}{\xi}.$$

Taking $h = 1$ and $g = \eta + \xi^k$ we have

$$\text{(iv)} \qquad \frac{d\eta}{\xi^k} + \frac{kd\xi}{\xi}.$$

Taking linear combinations of (iii) and (iv) we see that we can reach all of

$$\text{(v)} \qquad \mathcal{O}_{Y,x}\frac{d\xi}{\xi}, \mathcal{O}_{Y,x}\frac{d\eta}{\xi^k}$$

for any k. It remains to show that we can reach

$$\frac{\frac{d\xi}{\xi^{a+1}}}{\mathcal{O}_X \xrightarrow{\eta} \mathcal{O}_X}$$

and

$$\frac{\frac{d\xi}{\xi^{a+1}}}{\mathcal{O}_X \xrightarrow{\eta^b} \mathcal{O}_X}$$

for all positive integers a, b.

Now we come to the essential point, which is to *consider arcs in the space of cycles*. Thus, suppose we have an arc

$$\text{(vi)} \qquad (fk - th, g)$$

in $(\bigoplus_Y \underline{\underline{\mathbb{C}}}(Y)^*)_x$. Here, f and k are relatively prime in $\mathcal{O}_{X,x}$, $h \in \mathcal{O}_{X,x}$, and $g \in \mathbb{C}(X)$ does not have polar locus containing any component of $fk - th = 0$. All of this is local in the Zariski topology. What is the tangent in $\underset{Y}{\oplus} H^1_y(\Omega^1_{X/\mathbb{Q}})_x$ to the arc (vi)? Letting $Y_1 = \{f = 0\}$, $Y_2 = \{k = 0\}$, on the face of it our prescription above gives something like

$$\frac{h\frac{dg}{g}}{\mathcal{O}_X \xrightarrow{fk} \mathcal{O}_X}.$$

But we need to have something in the direct sum

$$\underline{\underline{H}}^1_{y_1}(\Omega^1_{X/\mathbb{Q}})_x \oplus \underline{\underline{H}}^1_{y_2}(\Omega^1_{X/\mathbb{Q}})_x.$$

We *define* this to be

$$\text{(vii)} \qquad \frac{\frac{h}{k}\frac{dg}{g}}{\mathcal{O}_X \xrightarrow{f} \mathcal{O}_X} + \frac{\frac{h}{f}\frac{dg}{g}}{\mathcal{O}_X \xrightarrow{k} \mathcal{O}_X}.$$

It is necessary to verify that this definition makes sense. For example, if $h = ku$ then as arcs in $Z^1_1(X)$

$$\begin{aligned}(fk - th, g) &= (fk - tuk, g)\\ &= (f - tu, g) + (k, g)\end{aligned}$$

and the second term is constant so that (vii) reduces to

$$\frac{\frac{h}{k}\frac{dg}{g}}{\mathcal{O}_X \xrightarrow{f} \mathcal{O}_X}.$$

We can also check that this definition is compatible with the description above using differential forms.

With this understood we have

$$\text{(viii)} \qquad (\xi\eta - t, g) \to \frac{\frac{dg}{\xi g}}{\mathcal{O}_X \xrightarrow{\eta} \mathcal{O}_X} + \frac{\frac{dg}{\eta g}}{\mathcal{O}_X \xrightarrow{\xi} \mathcal{O}_X}.$$

Now take $g = \xi^m + \eta^n$. Then the first term is

$$\frac{\frac{m\xi^{m-1}d\xi}{\xi(\xi^m+\eta^n)}}{\mathcal{O}_X \xrightarrow{\eta} \mathcal{O}_X} + \frac{\frac{n\eta^{n-1}d\eta}{\xi(\xi^m+\eta^n)}}{\mathcal{O}_X \xrightarrow{\eta} \mathcal{O}_X}.$$

Write

$$\frac{1}{\xi^m + \eta^n} - \frac{1}{\xi^m} = \frac{-\eta^n}{\xi^m + \eta^n} = \left(\frac{-\eta^{n-1}}{\xi^m + \eta^n}\right)\eta.$$

Since we mod out by $\eta\psi$'s where ψ does not have polar locus containing Y we may replace the expression above by

$$\frac{md\xi}{\xi^2}$$

and then in (viii)

$$(\xi\eta - t, \xi^m + \eta^n) \to \frac{\frac{md\xi}{\xi^2}}{\mathcal{O}_X \xrightarrow{\eta} \mathcal{O}_X} + \frac{\frac{nd\eta}{\eta^2}}{\mathcal{O}_X \xrightarrow{\xi} \mathcal{O}_X}.$$

Taking linear combinations with different m and n we may reach

$$\frac{\frac{d\xi}{\xi^2}}{\mathcal{O}_X \xrightarrow{\eta} \mathcal{O}_X}.$$

Replacing $\xi\eta - t$ by $\xi\eta - th$ we may then reach $\mathcal{O}_{Y,x}$ times this form in the numerator.

Similar considerations give

$$(\xi^a\eta - t, \xi^{an} + \eta^n) \to \frac{\frac{amd\xi}{\xi^{a+1}}}{\mathcal{O}_X \xrightarrow{\eta} \mathcal{O}_X} + \frac{\frac{nd\eta}{\eta^2}}{\mathcal{O}_X \xrightarrow{\xi^a} \mathcal{O}_X}$$

modulo terms already reached. Recalling that we are taking the limit over powers of the defining equations, and that by the previous step we may reach $\frac{nd\eta/\eta^2}{\mathcal{O}_X \xrightarrow{\xi} \mathcal{O}_X}$, we may reach the second term, and again taking linear combinations we may reach

$$\frac{\frac{d\xi}{\xi^{a+1}}}{\mathcal{O}_X \xrightarrow{\eta} \mathcal{O}_X}.$$

Finally, working modulo terms already reached we have

$$\left(\xi^a\eta^b - t, \xi^{am} + \eta^{bn}\right) \longrightarrow \frac{\frac{d\xi}{\xi^{a+1}}}{\mathcal{O}_X \xrightarrow{\eta^b} \mathcal{O}_X}.$$

Throughout we may replace t by ht with the effect of multiplying everything by $\mathcal{O}_{Y,x}$. This completes the proof of step two.

For step three we let $\widetilde{Y} \xrightarrow{\pi} Y \subset X$ be the normalization of the irreducible curve Y, and we let ξ, η be local uniformizing parameters centered at x and such that the projection $\xi, \eta \to \xi$ realizes Y as a branched covering over the curve $\eta = 0$ (all of this being local in the Zariski topology). The essential observation is this:

(iv) *The map*

$$k \to \pi^*\left(\frac{k}{\xi^a}\right) \in \mathbb{C}(\widetilde{Y}), \qquad k \in \mathcal{O}_{X,x} \text{ and } a \in \mathbb{Z}$$

is surjective.

We now consider arcs of the form $(f^a\xi^b - th, g)$ with tangent

$$\frac{\frac{h}{g}\left(\frac{da}{\xi^b}\right)}{\mathcal{O}_X \xrightarrow{f^a} \mathcal{O}_X} + \frac{\frac{h}{g}\left(\frac{dg}{f^a}\right)}{\mathcal{O}_X \xrightarrow{\xi^b} \mathcal{O}_X}.$$

By step two the second term may be reached by an arc in $\bigoplus_Y \mathbb{C}(Y)^*$. Subtracting this off and using (iv) above we may reach all of

$$\mathbb{C}(\widetilde{Y}) \otimes_{\mathbb{C}} \Omega^1_{X/\mathbb{Q},x}\big|_x. \qquad \square$$

8.4 IDENTIFICATION OF THE GEOMETRIC AND FORMAL TANGENT SPACES TO $CH^2(X)$ FOR X A SURFACE

Above we have defined

$$TZ^2(X) = H^0\left(\bigoplus_{x\in X} \underline{\underline{H}}^2_x\left(\Omega^1_{X/\mathbb{Q}}\right)\right) \tag{8.41}$$

and

$$TZ^1_1(X) = H^0\left(\bigoplus_{\left\{\substack{Y \text{ irred} \\ \operatorname{codim} Y=1}\right.} \underline{\underline{H}}^1_y\left(\Omega^1_{X/\mathbb{Q}}\right)\right) \tag{8.42}$$

Moreover, there are maps

$$\begin{array}{ccc} Z^1_1(X) & \longrightarrow & Z^2(X) \\ \| & & \| \\ \bigoplus_{\left\{\substack{Y \text{ irred} \\ \operatorname{codim} Y=1}\right.} \underline{\underline{\mathbb{C}}}(Y)^* & \xrightarrow{\text{div}} & \bigoplus_{x\in X} \underline{\underline{\mathbb{Z}}}_x \end{array} \tag{8.43}$$

and

$$TZ^1_1(X) \xrightarrow{\text{res}} TZ^2(X) \tag{8.44}$$

where using the isomorphisms in (8.41) and (8.42) the map (8.44) is given by the map in the Cousin flasque resolution of $\Omega^1_{X/\mathbb{Q}}$. Moreover, and this is a central point,

(8.45) *the mapping (8.44) is the differential of (8.43).*

We set

$$Z^2_{\text{rat}}(X) = \text{image of } \left\{Z^1_1(X) \to Z^2(X)\right\}$$

and recall that the *Chow group* is defined by

$$CH^2(X) = Z^2(X)/Z^2_{\text{rat}}(X).$$

Definition: *The geometric tangent space to $CH^2(X)$ is defined by*

$$T_{\text{geom}}CH^2(X) = TZ^2(X)/\text{image}\left\{TZ^1_1(X) \xrightarrow{\text{res}} TZ^2(X)\right\}. \tag{8.46}$$

The word "geometric" refers to the fact that the right-hand side is given by geometric data: the numerator by tangents to arcs in $Z^2(X)$ and the denominator by the tangents to image of arcs in $Z^1_1(X)$ under the map "div."

A tangent vector in $T_{\text{geom}}CH^2(X)$ is thus represented by data

$$\tau = \sum_i \tau_i,$$

where there are points $x_i \in X$ and where, using duality to express local cohomology by differential forms,

$$\tau_i \in \text{Hom}^o\left(\Omega^2_{X/\mathbb{Q},x_i}/\Omega^2_{\mathbb{C}/\mathbb{Q}}, \Omega^1_{\mathbb{C}/\mathbb{Q}}\right).$$

A vector in the subspace image $\left\{TZ_1^1(X) \to TZ^2(X)\right\}$ is represented by the residue of data

$$\varphi = \sum_\nu \varphi_\nu$$

where there are irreducible curves $Y_\nu \subset X$ and

$$\varphi_\nu \in \mathrm{Hom}_{\mathcal{O}_Y}\left(\Omega^2_{X/\mathbb{Q},Y_\nu}/\Omega^2_{\mathbb{C}/\mathbb{Q}}, \Omega^1_{\mathbb{C}(Y_\nu)/\mathbb{C}} \otimes \Omega^1_{\mathbb{C}/\mathbb{Q}}\right).$$

Both the data $\{\tau_i\}$ and $\{\varphi_\nu\}$ are geometric in character.

Remark that although (8.46) is geometric it is definitely "nonclassical." For example, as will be discussed in chapter 10 below, there are algebraic arcs $z(t)$ in $Z^2(X)$ such that for all t

$$z'(t) \in TZ^2_{\mathrm{rat}}(X),$$

where $TZ^2_{\mathrm{rat}}(X)$ denotes the denominator on the RHS of (8.46), but where

$$z(t) \text{ is nonconstant in } CH^2(X).$$

Thus, for what will seem to be arithmetic reasons, $Z^2_{\mathrm{rat}}(X)$ is definitely *not* a "subvariety" of $Z^2(X)$ in anything resembling the usual sense.

Many years ago, Spencer Bloch established the identification

$$CH^2(X) \cong H^2(\mathcal{K}_2(\mathcal{O}_X))$$

and on the basis of this together with the van der Kallen isomorphism

$$T\mathcal{K}_2(\mathcal{O}_X) \cong \Omega^1_{X/\mathbb{Q}}$$

defined what we shall call the *formal tangent space*

$$T_{\mathrm{formal}}CH^2(X) = H^2\left(\Omega^1_{X/\mathbb{Q}}\right).$$

The main result in this work is the following

(8.47) **Theorem:** *There is a natural isomorphism*

$$T_{\mathrm{formal}}CH^2(X) \cong T_{\mathrm{geom}}CH^2(X).$$

The proof of this result is a consequence of Theorem (8.22) above together with the Cousin flasque resolution

$$(8.48)\quad 0 \to \Omega^1_{X/\mathbb{Q}} \to \Omega^1_{\mathbb{C}(X)/\mathbb{Q}} \to \bigoplus_{\left\{\substack{Y \text{ irred} \\ \operatorname{codim} Y=1}\right.} \underline{\underline{H}}^1_y\left(\Omega^1_{X/\mathbb{Q}}\right) \to \bigoplus_{x\in X} \underline{\underline{H}}^2_x\left(\Omega^1_{X/\mathbb{Q}}\right) \to 0$$

discussed in section 8.1.1.

Remarks: (i) The result in Theorem (8.47) implies an *existence theorem*, albeit at the infinitesimal level. Namely, given data $\tau \in \sum_i \tau_i$ as above, we may consider that

$$\tau \in T_{\mathrm{formal}}CH^2(X) = H^2\left(\Omega^1_{X/\mathbb{Q}}\right).$$

If in this interpretation $\tau = 0$, then there exists a global configuration $\{Y_\nu, \varphi_\nu\}$ consisting of irreducible curves Y_ν and data φ_ν on Y_ν that under the residue construction maps to τ.

This may be thought of as somewhat analogous to a 2-dimensional version of the following result for a smooth curve Y: Let $y_i \in Y$ and $\tau_i \in \mathcal{PP}_{Y,y_i}$ be Laurent tails that satisfiy

$$\sum_i \mathrm{Res}_{y_i}(\tau_i \omega) = 0$$

for all $\omega \in H^0(\Omega^1_{Y/\mathbb{C}})$. Then there exists a global rational function $f \in \mathbb{C}(Y)^*$ with

$$\text{principal part of } f \text{ at } y_i = \tau_i.$$

(ii) The theorem has the following consequence, which is an infinitesimal analogue of a conjecture of Bloch and Beilinson:

(8.49) **Corollary:** *Let X be an algebraic surface defined over a number field, and assume given tangent vectors*

$$\tau_i \in T_{x_i} X$$

such that (a) for all $\varphi \in H^0\left(\Omega^1_{X/\mathbb{C}}\right)$

$$\sum_i \langle \varphi, \tau_i \rangle = 0$$

and (b) the points $x_i \in X(\bar{\mathbb{Q}})$. Then there exists an infinitesimal rational equivalence given by irreducible curves Y_ν and data φ_ν on Y_ν that cuts out $\tau = \sum_i \tau_i$ in the sense that

$$\sum_\nu \mathrm{Res}(\varphi_\nu) = \sum_i \tau_i .$$

We may also take the (Y_ν, φ_ν) to be defined over $\bar{\mathbb{Q}}$.

Condition (a) is equivalent to $\sum_i (X_i, \tau_i) \in TX$ being in the kernel of the differential of the Albanese map.

A variant of this is the following

(8.50) **Corollary:** *Let X be as above and assume that X is regular. Assume that the $x_i \in X(k)$ are defined over a field k and let*

$$\begin{aligned} &\mathcal{X} = X \times S \\ &\cup \\ &\mathcal{X}_i = k\text{-spread of } x_i, \end{aligned}$$

where $\mathbb{Q}(S) \cong k$ (cf. section 4.2 for the notation). Each $\omega \in H^0(\Omega^2_{X(\mathbb{Q})/\mathbb{Q}})$ defines a 2-form $\widetilde{\omega}$ on $\mathcal{X}$ and we assume that

$$\sum_i \tau_i \rfloor \widetilde{\omega} \in \bigoplus_i T^* \mathcal{X}_i$$

is zero. Then $\tau = \sum_i \tau_i$ is tangent to an infinitesimal rational equivalence as in Corollary (8.49).

(iii) As suggested earlier, it is reasonable to hope that it will be possible to define $TZ^p(X)$ in general and that the analogue of Theorem (8.47) will be valid. However, the analogue of the following consequence of (8.47),

$$\text{every } \tau \in T_{\text{formal}} CH^2(X) \text{ is tangent to an arc in } Z^2(X),$$

will in general be false. This phenomenon, which may be related to the phenomenon noted above that $Z^2_{\text{rat}}(X)$ is not a "subvariety" of $Z^2(X)$, will be discussed in chapter 10 below. Remark that the principle seems to be the following:

There are formal constructions of compatible maps

$$\begin{cases} \varphi_n : \mathbb{C}[t]/t^{n+1} \to Z^p(X), \\ \varphi_n : \mathbb{C}[t]/t^{n+1} \to Z^p_{\text{rat}}(X) \end{cases}$$

for $n = 1, 2, \dots$ *for which no convergent constructions exist with a given tangent* φ_1.

This is in contrast to classical algebraic geometry, where the principle

$$\text{formal} \Rightarrow \text{actual}$$

holds. The reason for this nonclassical phenomenon seems to be a subtle mix of Hodge theory and/or arithmetic (cf. chapter 10).

To conclude this section we will discuss how to calculate the differential of the map

$$CH^1(Y) \xrightarrow{i} CH^2(X), \tag{8.51}$$

where $Y \subset X$ is a curve with normal crossings. In fact, the exact cohomology sequence of the exact sheaf sequence

$$0 \to \Omega^1_{X/\mathbb{Q}} \to \Omega^1_{X/\mathbb{Q}}(\log Y) \xrightarrow{\text{Res}} \mathcal{O}_Y \to 0 \tag{8.52}$$

gives a map

$$H^1(\mathcal{O}_Y) \xrightarrow{\delta} H^2\left(\Omega^1_{X/\mathbb{Q}}\right).$$

Observation: *The diagram*

$$\begin{array}{ccc} TCH^1(Y) & \xrightarrow{i_*} & TCH^2(X) \\ \wr\| & & \wr\| \\ H^1(\mathcal{O}_Y) & \xrightarrow{\delta} & H^2\left(\Omega^1_{X/\mathbb{Q}}\right) \end{array}$$

commutes, where the vertical isomorphisms are the natural identifications discussed above and i_* *is the differential of the map (8.51).*

In fact, we can say a little more. The extension class of (8.51) lies in

$$\mathcal{E}xt^1_{\mathcal{O}_X}\left(\mathcal{O}_Y, \Omega^1_{X/\mathbb{Q}}\right)$$

and maps to $H^1_Y\left(\Omega^1_{X/\mathbb{Q}}\right)$ to define the *local arithmetic cycle class*

$$[Y]_{a,\text{loc}} \in H^1_Y\left(\Omega^1_{X/\mathbb{Q}}\right).$$

In this sense we have that

$$[Y]_{a,\text{loc}} \textit{ determines } TCH^1(Y) \xrightarrow{i_*} TCH^2(X).$$

8.5 CANONICAL FILTRATION ON $TCH^n(X)$ AND ITS RELATION TO THE CONJECTURAL FILTRATION ON $CH^n(X)$

Beilinson has conjectured the existence of a canonical filtration $F^m CH^p(X)$ on the Chow groups $CH^p(X)$ of a smooth projective variety and, assuming the existence of the category of mixed motives, has proposed a Hodge-theoretic interpretation of the associated graded of this filtration.[7] H. Saito-Jannsen [25] and [28] and Murre [29] among others have proposed definitions for $F^m CH^p(X)$. In this section we shall show that

(i) there is a canonical filtration $F^m TCH^n(X)$ such that $Gr^m TCH(S)$ has a Hodge-theoretic description;

(ii) both the H. Saito-Jannsen and Murre definitions satisfy

$$TF^m CH^n(X) \subseteq F^{m-1} TCH^n(X);^8$$

and

(iii) there is a geometric interpretation of $F^m TCH^n(X)$ in terms of the transcendence level of cycles representing a class in $TCH^n(X)$.

We begin with step (i). It is based on the extension of the identification in Theorem (8.47) to n-dimensional smooth varieties to give the isomorphism

$$TCH^p(X) \cong H^p\left(\Omega^{p-1}_{X/\mathbb{Q}}\right). \tag{8.53}$$

For all p there is a canonical filtration on $\Omega^p_{X/\mathbb{Q}}$ given by

$$F^m \Omega^p_{X/\mathbb{Q}} = \text{image of } \left\{\Omega^m_{\mathbb{C}/\mathbb{Q}} \otimes \Omega^{p-m}_{X/\mathbb{Q}} \to \Omega^p_{X/\mathbb{Q}}\right\}. \tag{8.54}$$

The associated graded of (8.54) is

$$Gr^m \Omega^p_{X/\mathbb{Q}} = \Omega^m_{\mathbb{C}/\mathbb{Q}} \otimes \Omega^{p-m}_{X/\mathbb{C}}.$$

Geometrically, we may think of this filtration as reflecting the Leray filtration along X in the spread

$$\mathcal{X} \to S$$

of X (cf. section 4.2). The *arithmetic Gauss Manin connection*

$$\nabla_{X/\mathbb{Q}} : \Omega^m_{\mathbb{C}/\mathbb{Q}} \otimes H^q\left(\Omega^{p-m}_{X/\mathbb{C}}\right) \to \Omega^{m+1}_{\mathbb{C}/\mathbb{Q}} \otimes H^{q+1}\left(\Omega^{p-m-1}_{X/\mathbb{C}}\right) \tag{8.55}$$

is induced from the extension classes of the filtration (8.54); it may be thought of as the usual Gauss-Manin connection for the spread looked at along X. With this interpretation it was proved by Esnault-Paranjape [18] that the spectral sequence associated to (8.54) degenerates at E_2. From this it follows that the induced filtration on $H^q(\Omega^p_{X/\mathbb{Q}})$ is given by

$$F^m H^q\left(\Omega^p_{X/\mathbb{Q}}\right) = \text{image of } \left\{H^q\left(F^m \Omega^p_{X/\mathbb{Q}}\right) \to H^q\left(\Omega^p_{X/\mathbb{Q}}\right)\right\}, \tag{8.56}$$

---

[7] The discussion in this section is modulo torsion.

[8] This discussion will work for the formal tangent space $TCH^p(X)$ for all p, and thus will work in general if $TZ^p(X)$ can be defined so that the analogue of (8.47) holds.

and the associated graded is

$$Gr^m H^q\left(\Omega^p_{X/\mathbb{Q}}\right) = \left(\Omega^m_{\mathbb{C}/\mathbb{Q}} \otimes H^q\left(\Omega^{p-m}_{X/\mathbb{C}}\right)\right)_{\nabla}, \tag{8.57}$$

where ∇ is $\nabla_{X/\mathbb{Q}}$ and the right-hand side is the cohomology of the complex arising from (8.55) at the indicated spot.

Definition. *With the identification (8.53), the canonical filtration on $TCH^p(X)$ is defined by*

$$F^m TCH^p(X) = F^{m-1}H^n\left(\Omega^{p-1}_{X/\mathbb{Q}}\right). \tag{8.58}$$

Remarks: For the conjectural filtration on $CH^n(X)$, there is agreement that

$$\begin{aligned} F^1CH^p(X) &= \ker\left\{CH^p(X) \to H^{2p}(X,\mathbb{Z})\right\}, \\ F^2CH^p(X) &= \ker\left\{F^1CH^p(X) \xrightarrow{AJ^p_X} J^p(X)\right\}, \end{aligned} \tag{8.59}$$

where $J^p(X)$ is the intermediate Jacobian variety of X and AJ^p_X is the Abel-Jacobi mapping. Since $H^{2p}(X,\mathbb{Z})$ is discrete and therefore does not show up in infinitesimal considerations, the filtration (8.58) is shifted by one in order to have the possibility that

$$F^m TCH^p(X) = TF^mCH^p(X).$$

We note that by (8.59)

$$\begin{aligned} F^2TCH^p(X) &= \ker dAJ^p_X \\ &= TF^2CH^p(X), \end{aligned}$$

where dAJ^p_X is the differential of the Albanese mapping.

We now turn to (ii).

Proposition: *Denoting by $F^mCH^p(X)$ the filtration on $CH^p(X)$ defined by H. Saito-Jannsen or by Murre, we have*

$$TF^mCH^p(X) \subseteq F^mTCH^p(X). \tag{8.60}$$

Proof: The H. Saito-Jannsen definition of $F^mCH^p(X)_{\mathbb{Q}}$ is as follows:[9]

> $F^mCH^n(X)_{\mathbb{Q}}$ is generated by cycles of the form
>
> $$(p_X)_*(Z_1\cdots Z_m)$$
>
> where, for some parameter variety T,
>
> $$\begin{cases} Z_i \in F^1CH^{q_i}(X\times T) = CH^{q_i}(X\times T)_{\text{hom}}, \\ Z_i \equiv_{\text{hom}} 0. \end{cases}$$

---

[9] Here, for any $\mathbb{Z}$-module M we are setting $M_{\mathbb{Q}} = M \otimes_{\mathbb{Z}} \mathbb{Q}$. Also, H. Saito-Jannsen give the (same) definition of $F^mCH^p(X)$ for all p.

To obtain the tangent to an arc in $F^m CH^p(X)$, using the evident notation we have

$$\dot{Z}_1 \in H^{q_1}\left(\Omega^{q_1-1}_{X\times T/\mathbb{Q}}\right)$$

and

$$[Z_i]_a \in \text{image}\left\{H^{q_i}\left(F^1\Omega^{q_i}_{X\times T/\mathbb{Q}}\right) \to H^{q_i}\left(\Omega^{q_i}_{X\times T/\mathbb{Q}}\right)\right\}.$$

Consequently

$$\dot{Z}_1 \cup [Z_2]_a \cup \cdots \cup [Z_m]_a \in \text{image}\left\{H^{q_1+\cdots+q_m}\left(F^{m-1}\Omega^{q_1+\cdots+q_m-1}_{X\times T/\mathbb{Q}}\right) \right.$$
$$\left. \to H^{q_1+\cdots+q_m}\left(\Omega^{q_1+\cdots+q_m-1}_{X\times T/\mathbb{Q}}\right)\right\}.$$

Applying $(p_X)_*$ we end up in

$$\text{image}\left\{H^p\left(F^{m-1}\Omega^{n-1}_{X/\mathbb{Q}}\right) \to H^p\left(\Omega^{n-1}_{X/\mathbb{Q}}\right)\right\}$$

as desired.

Murre's conjectural filtration on $CH^p(X)_\mathbb{Q}$ is constructed inductively as follows (again his construction works for all $CH^p(X)_\mathbb{Q}$), Denoting, as usual, the diagonal by $\Delta \subset X \times X$ with fundamental class

$$[\Delta] \in H^{2p}(X \times X)_\mathbb{Q} \cong \bigoplus_{k=0}^{2p} \text{Hom}\left(H^k(X)_\mathbb{Q}, H^k(X)_\mathbb{Q}\right).$$

The kth Künnuth component is a Hodge class in $H^{2p}(X \times X)_\mathbb{Q}$, and—assuming the Hodge conjecture—is represented by a cycle Δ_k which induces a map

$$\Delta_k : CH^p(X)_\mathbb{Q} \to CH^p(X)_\mathbb{Q}.$$

Now

$$\Delta_{2p} = \text{identity on } Gr^0 CH^p(X)_\mathbb{Q}$$

and we set

$$F^1 CH^p(X)_\mathbb{Q} = \ker \Delta_{2p}.$$

Next,

$$\Delta_{2p-1} = \text{identity on } Gr^1 CH^p(X)_\mathbb{Q}$$

and we set

$$F^2 CH^p(X)_\mathbb{Q} = \ker\left(\Delta_{2p-1}\Big|_{\ker(\Delta_{2p})}\right),$$

and so forth.

To prove (8.60) for the Murre filtration, we assume that we have an arc $z(t)$ in $Z^n(X)$ given by

$$\mathcal{Z} \subset X \times B.$$

We assume that

$$(p_2)_*\left(p_1^* z(t) \cdot [\Delta_k]\right) \equiv_{\text{rat}} 0$$

for $t \in B$, where p_i $(i = 1, 2)$ are the coordinate projections $X \times X \to X_i$. Now

$$[\mathcal{Z}]_a \in H^k\left(\Omega^n_{X\times B/\mathbb{Q}}\right),$$
$$[\Delta]_a \in H^n\left(\Omega^n_{X\times X/\mathbb{Q}}\right)$$

and, as explained in chapter 7,

$$z'(0) \in H^n\left(\Omega^{n-1}_{X/\mathbb{Q}}\right) \otimes \Omega^1_{B/\mathbb{C}}$$

represents part of $[\mathcal{Z}]_a$. Thus

$$(p_2)_*\left(p_1^*(z'(0)) \cdot [\Delta_k]\right) = 0.$$

Now the topological part of $[\Delta_k]$ is

$$\mathrm{id}_{H^k(X)} \in H^n\left(\Omega^n_{X\times X/\mathbb{C}}\right),$$

and

$$\alpha \to (p_2)_*\left(p_1^*\alpha \cdot [\Delta_k]\right)$$

induces the identity on

$$Gr^{2n-1-k} H^n\left(\Omega^{n-1}_{X/\mathbb{Q}}\right) \cong \left(\Omega^{2n-1-k}_{\mathbb{C}/\mathbb{Q}} \otimes H^n\left(\Omega^{n-k}\right)\right)_\Delta.$$

Thus we can tell which $F^m H^n\left(\Omega^{n-1}_{X/\mathbb{Q}}\right)$ the class $z'(0)$ belongs to by which Δ_k annihilates it. The indices work out to give (8.60). □

Turning to the third point, in order to illustrate the essential geometric idea we shall assume that X is defined over $\mathbb{Q}$ (or over a number field). For $w = x_1 + \cdots + x_d \in X^{(d)}$ where the x_i are assumed distinct, we shall denote by

$$T_w Z^n(X) \subset TZ^n(X)$$

the tangents to arcs $z(t)$ in $Z^n(X)$ with

$$\lim_{t\to 0} |z(t)| = w.$$

There is then a natural mapping

$$T_w Z^n(X) \to TCH^n(X)$$

induced from the geometric mapping

$$TZ^n(X) \to TCH^n(X)$$

discussed in section 8.3. From that discussion follows a proof of the

Proposition: *If* tr deg $w \leqq m$ *and* $\dot{z}(0) \in F^{m+1}TCH^n(X)$, *then* $\dot{z}(0) = 0$ *in* $TCH^n(X)$. *For example, if* X and w are defined over $\bar{\mathbb{Q}}$ *and* $\dot{z}(0) \in F^1TCH^n(X)$, *i.e.* $\dot{z}(0) = 0$ *in* $H^n\left(\Omega^{n-1}_{X/\mathbb{C}}\right)$, *then* $\dot{z}(0) = 0$ *in* $H^n\left(\Omega^{n-1}_{X/\mathbb{Q}}\right)$.

In other words, infinitesimally the filtration on $CH^n(X)$ reflects the transcendence level of cycles representing classes in the Chow groups. In [32] we have proposed yet another definition of $F^m CH^p(X)$ for all p, together with a set of Hodge-theoretic invariants of $Gr^m CH^p(X)$ that will—assuming the conjectures of Bloch-Beilinson—capture rational equivalence modulo torsion. Our proposed definition of $F^m CH^p(X)$ will be compatible with those of H. Saito-Jannsen and Murre, but will differ from theirs in that our condition on a cycle Z that

$$[Z] \in F^m CH^p(X)$$

will be testable in finite terms. Its infinitesimal description is given by (8.58).

Chapter Nine

Applications and Examples

9.1 THE GENERALIZATION OF ABEL'S DIFFERENTIAL EQUATIONS

Let X be a smooth algebraic curve and

$$\tau = \sum_{i=1}^{d} (x_i, \tau_i)$$

a configuration of points $x_i \in X$ (assumed for simplicity of exposition to be distinct) and tangent vectors $\tau_i \in T_{x_i}X$. Classically Abel's differential equations (cf. [36]) dealt with the question

(9.1.) *When is τ tangent to a rational equivalence?*

The answer, which emerged from the work of Abel, Jacobi, Riemann and others in the nineteenth century is

The necessary and sufficient condition that τ be tangent to a rational equivalence is that for every regular differential $\omega \in H^0(\Omega^1_{X/\mathbb{C}})$

$$\langle \omega, \tau \rangle =: \sum_{i=1}^{d} \langle \omega(x_i), \tau_i \rangle = 0. \tag{9.2}$$

It is the higher dimensional analogue of this statement that we shall be discussing, focusing on the case of configurations of points on a surface.

We first reformulate (9.2). On the symmetric product $X^{(d)}$ we have the regular 1-forms

$$\operatorname{Tr} \omega \in H^0(\Omega^1_{X^{(d)}/\mathbb{C}})$$

defined for $\omega \in H^0(\Omega^1_{X/\mathbb{C}})$, and (9.2) is equivalent to

$$\langle \operatorname{Tr} \omega, \tau \rangle = 0 \text{ for } \tau \in T_z X^{(d)},$$

where $z = x_1 + \cdots + x_d$, and the x_i need not be distinct. The equations

$$\operatorname{Tr} \omega = 0 \text{ on } X^{(d)} \tag{9.3}$$

define an exterior differential system on $X^{(d)}$ which we call *Abel's differential equations on $X^{(d)}$*. Referring to [30] for an explanation of the terminology, it is a consequence of the theorems of Abel, Jacobi, and Riemann-Roch that:

(9.4) *Abel's differential equations constitute an involutive exterior differential system. The maximal integral manifold through $z \in X^{(d)}$ is given by the complete linear system*

$$|z| = \{z' \in X^{(d)} : z' - z = (f) \textit{ for some } f \in \mathbb{C}(X)^*\}$$

and

$$\dim |z| = g - d + i(z),$$

where $g = \dim H^0(\Omega^1_{X/\mathbb{C}})$ *and*

$$i(z) = \dim\{\tau \in T_z X^{(d)} : \langle \mathrm{Tr}\, \omega, \tau\rangle = 0 \textit{ for all } \omega \in H^o(\Omega^1_{X/\mathbb{C}})\}.$$

Before taking up the situation of 0-cycles on a surface we want to relate (9.4) to the topic of this book, which is the tangent space to the space of cycles. Namely, we have

$$TZ^1(X) \cong \bigoplus_{x \in X} \mathrm{Hom}^c_{\mathbb{C}}(\Omega^1_{X/\mathbb{C},x}, \mathbb{C}).$$

Each regular differential $\omega \in H^0(\Omega^1_{X/\mathbb{C}})$ gives in the obvious way

$$\mathrm{Tr}\, \omega \in T^* Z^1(X).$$

By definition, Abel's differential equations are given by

$$\mathrm{Tr}\, \omega = 0 \text{ on } Z^1(X). \tag{9.5}$$

The difference between (9.3) and (9.5) is that *in (9.5) creation and annihilation are allowed.* That is, for z and $z' \in X^{(d)}$ a rational equivalence between z and z' will in principle allow the existence of $w \in X^{(d')}$ and a map

$$\mathbb{P}^1 \xrightarrow{f} X^{(d+d')}$$

with

$$\begin{cases} f(0) = z + w, \\ f(\infty) = z' + w. \end{cases}$$

For curves, this is of course not necessary, which we may formulate by considering the obvious map

$$X^{(d)} \xrightarrow{\pi} Z^1(X)$$

and saying that for $z \in X^{(d)}$ the integral manifold of (9.5) through $\pi(z)$ lies in $\pi(X^{(d)})$. In this sense, for algebraic curves there is no real need to consider $Z^1(X)$, other than of course that it makes the formalism neater. This situation will change completely when considering higher codimension cycles.

We now turn to (9.1) on an algebraic surface. For reasons to be discussed below the question needs to be refined into three parts:

(9.6) (a) When is τ tangent to a first order rational equivalence?

(9.6) (b) When is τ tangent to a formal rational equivalence?

(9.6) (c) When is τ tangent to a geometric rational equivalence?

Here, (a) has the following meaning: τ defines a tangent vector, still denoted by τ, in

$$\underline{\underline{T}}Z^2(X) =: \bigoplus_{x\in X} \lim_{k\to\infty} \mathcal{E}xt^2_{\mathcal{O}_X}\left(\mathcal{O}_X/\mathfrak{m}_x^k, \Omega^1_{X/\mathbb{Q}}\right)$$
$$\cong \bigoplus_{x\in X} \mathrm{Hom}^o(\Omega^2_{X/\mathbb{Q},x}, /\Omega^2_{\mathbb{C}/\mathbb{Q}}, \Omega^1_{\mathbb{C}/\mathbb{Q}}).$$

From chapter 8 we have a natural map

$$T\Big(\bigoplus_{\left\{\substack{Y \text{ codim } 1\\ Y \text{ irred}}\right.} \mathbb{C}(Y)^*\Big) \xrightarrow{\rho} TZ^2(X),$$

and (a) means that τ should be in the image of ρ. As for (b), there should be a formal arc $z(t)$ in $Z^2(X)$ and a formal arc $\xi(t)$ in $\bigoplus_Y \mathbb{C}(Y)^*$ such that

$$\rho\big(\xi'(0)\big) = z'(0) = \tau$$

and

$$\mathrm{div}\big(\xi(t)\big) = z(t) - z(0).$$

(A formal arc in $\bigoplus_Y \mathbb{C}(Y)^*$ means a map from $\lim_k \mathrm{Spec}(\mathbb{C}[\epsilon]/\epsilon^k)$ into this space.) Finally, (c) means the same statement dropping the "formal"—there should be a geometric cycle in $B \times X \times \mathbb{P}^1$ and $t_0 \in B$ such that the induced map $T_{t_0}B \to TZ^1_1(X)$ has τ in its image. From the discussion in chapter 8, (a) is a linear cohomological question which we shall now discuss, postponing (b) and (c) to the next section.

We will first discuss (a) informally without complete proofs, emphasizing some consequences of the result. Then we shall discuss it more formally, giving proofs and tying the situation into the interpretation of $\mathcal{E}xt^2$ in terms of fourth quarter Laurent tails that was given at the end of chapter 7.

Let us first assume that X is defined over $\mathbb{Q}$. For $\tau \in TZ^2(X)$ and $\omega \in H^0(\Omega^2_{X/\mathbb{Q}})$ we have in chapter 5 defined

$$\langle \omega, \tau \rangle \in \Omega^1_{\mathbb{C}/\mathbb{Q}}$$

with the property that, if $\omega = \varphi \wedge \alpha$ where $\varphi \in H^0(\Omega^1_{X/\mathbb{Q}})$ and $\alpha \in \Omega^1_{\mathbb{C}/\mathbb{Q}}$, then

$$\langle \varphi \wedge \alpha, \tau \rangle = \langle \varphi, \tau \rangle \alpha,$$

where $\langle \varphi, \tau \rangle$ is the usual pairing of 1-forms against tangent vectors. We observe that this expression depends only on the image of φ in $H^0(\Omega^1_{X/\mathbb{C}})$, so that we may intuitively think of $\varphi \in H^0(\Omega^1_{X/\mathbb{C}})$. From chapter 8 we know that $TZ^2_{\mathrm{rat}}(X) \subset TZ^2(X)$ is defined by the conditions

$$\langle \omega, \tau \rangle = 0 \text{ for all } \omega \in H^0(\Omega^2_{X/\mathbb{Q}}), \tag{9.7}$$

which we shall call Abel's differential equations and write as

$$\mathrm{Tr}\,\omega = 0 \tag{9.8}$$

in analogy to (9.5).

To get some feeling for these equations, suppose that $\tau = \sum_{i=1}^{d}(x_i, \tau_i)$, where the x_i are distinct and $\tau_i \in T_{x_i}X$. Let $\xi, \eta \in \mathbb{Q}(X)$ give local uniformizing parameters around each x_i. If $\omega = f(\xi, \eta)d\xi \wedge d\eta$, where $f(\xi, \eta)$ is an algebraic function of ξ, η, then (9.7) is

$$\sum_i f(\xi_i, \eta_i)(d\xi_i \mu_i - d\eta_i \lambda_i) = 0, \tag{9.9}$$

where

$$\begin{cases} \tau_i = \lambda_i \partial/\partial\xi + \mu_i \partial/\partial\eta, \\ \xi_i = \xi(x_i),\ \eta_i = \eta(x_i). \end{cases}$$

Varying τ_i we see that *the rank of the equations (9.9) depends on the geometric position of the points x_i and on the field of definition of the x_i.* This mixture of geometry and arithmetic is characteristic of higher codimension cycles.

To give a simply stated corollary, we set $z = x_1 + \cdots + x_d$ and define

$$\operatorname{Tr} \deg z = \dim \operatorname{span}\{d\xi_i, d\eta_i\},$$

where $d = d_{\mathbb{C}/\mathbb{Q}}$. Then

$$0 \leqq \operatorname{Tr} \deg z \leqq 2d$$

with $\operatorname{Tr} \deg z = 0$ if, and only if, all the $x_i \in X(\bar{\mathbb{Q}})$, while $\operatorname{Tr} \deg z = 2d$ may be taken as a definition of z being *generic*.

Corollary: *For a surface X defined over $\bar{\mathbb{Q}}$ with $K_X \cong \mathcal{O}_X$ the rank of the equations (9.9) is equal to* Tr z, *where $\omega \in H^0(\Omega^2_{X/\mathbb{Q}})$ is a generator.*

It follows easily from the obvious extension of this corollary to the case when $H^0(\Omega^2_{X/\mathbb{C}}) \neq 0$ that there is no finite dimensional algebraic variety W together with a map $W \to Z^2(X)$ such that the induced mapping $W \to CH^2(X)$ is surjective. This type of result is due to Mumford, Roitman, and others; in all of their arguments the word "generic" is used to mean in the complement of a countable union of proper subvarieties. The above discussions gives a somewhat precise arithmetic-geometric meaning to what generic should mean (cf. the appendix to section 10.1 below).

At the other extreme, *if X defined over $\bar{\mathbb{Q}}$ and all the $x_i \in X(\bar{\mathbb{Q}})$ then the rank of the equations (9.9) is zero.* As explained in section 8.4 this result is an infinitesimal version of conjectures of Bloch and Beilinson. Namely, assuming that X is a regular surface, to say that the equations (9.9) are zero means that given *any* set of tangent vectors $\tau_i' \in T_{x_i}X$, *there exists* an arc $\{Y_\lambda(\epsilon), f_\lambda(\epsilon)\}$ $(\epsilon^2 = 0)$ where $Y_\lambda(\epsilon)$ is a family of irreducible curves and $f_\lambda(\epsilon) \in \mathbb{C}(Y_\lambda(\epsilon))^*$ with

$$\frac{d}{d\epsilon}\left(\sum_\lambda \operatorname{div} f_\lambda(\epsilon)\right) = \sum_i \tau_i.$$

The infinitesimal picture that we have developed has the following implications (here, X is still an algebraic surface defined over $\mathbb{Q}$ with $h^{2,0}(X) \neq 0$):

(i) *If* $z = \sum_{i=1}^{d} x_i$ *is generic, then it does not move in a rational equivalence.*

Here, "generic" means something both geometric and arithmetic. The geometric part is relative to the canonical linear system $|K_X|$, in analogy with nonspecial divisors in the theory of algebraic curves. The arithmetic part is that tr deg z should be sufficiently large—for example, tr deg $z = 2d$ will do.

(ii) *At the other extreme, if the* $x_i \in X(\bar{\mathbb{Q}})$ *then any infinitesimal motion of* z *is covered by a rational equivalence.*

The conjecture of Bloch-Beilinson implies that for any other $\tilde{z} = \sum_{i=1}^{d} \tilde{x}_i$ with $\tilde{x}_i \in X(\bar{\mathbb{Q}})$ there is a rational equivalence

$$z \equiv_{\text{rat}} \tilde{z}.$$

We may rephrase this by saying that on $X(\bar{\mathbb{Q}})$ the 0-cycle z moves in a rational equivalence class $|z|_{\bar{\mathbb{Q}}}$ where

$$\dim |z|_{\bar{\mathbb{Q}}} = 2d$$

Intermediate between (i) and (ii) we consider a fixed field k of transcendence degree $d(k)$. Then the infinitesimal theory suggests

(iii) *Any* $z = \sum_{i=1}^{d} x_i$, *where* $x_i \in X(k)$ *will move on* $X(k)$ *in a rational equivalence class* $|z|_k$, *where asymptotically*

$$\dim |z|_k \sim 2\big(d - d(k)\big).$$

Comparing (i)–(iii) suggests behavior analogous to linear equivalence on curves but where tr deg k replaces the genus.

Thus, for curves we have

$$\dim |z| \sim d - g + i(z),$$

where $z = \sum_{i=1}^{d} x_i$ is a divisor of degree d with complete linear system $|z|$ and g is the genus. The correction term $i(z)$ in this formula is the index of speciality, which for $d \leqq g$ measures the failure of the x_i to impose independent conditions on the canonical series. Geometrically, we have the canonical mapping

$$\varphi_K : X \to \mathbb{P}\big(H^1(\mathcal{O}_X)\big)$$

and $i(z)$ measures the failure of the $\varphi_K(x_i)$ to be in general position.

For a surface X as above we may define the *arithmetic canonical mapping*

$$\varphi_{K_a} : X(k) \to \mathbb{P}\big(H^2(\mathcal{O}_{X(k)}) \otimes \Omega^1_{k/\mathbb{Q}}\big),$$

and what is suggested is that

$$\dim |z|_k = 2d - 2d(k) - i(z),$$

where $i(z)$ is expressible in terms of the geometry of the points $\varphi_{K_a}(x_i)$.

We will now reformulate (9.7) in a way that will enable the extension of those equations to the case when X is not defined over $\mathbb{Q}$. Again the formal proof will be given later. First, remembering that we are still assuming that X is defined over $\mathbb{Q}$ the equations (9.7) include the equations

$$(9.7)_1 \qquad \langle \varphi, \tau \rangle = 0 \ \text{ for } \ \varphi \in H^0(\Omega^1_{X/\mathbb{C}}).$$

Next, since

$$H^0(\Omega^2_{X(\mathbb{Q})/\mathbb{Q}}) \otimes \mathbb{C} \cong H^0(\Omega^2_{X(\mathbb{C})/\mathbb{C}})$$

we may lift every $\omega \in H^0(\Omega^2_{X/\mathbb{C}})$ to $\widetilde{\omega} \in H^0(\Omega^2_{X/\mathbb{Q}})$; the lifting is unique up to $\Omega^1_{\mathbb{C}/\mathbb{Q}} \otimes H^0(\Omega^1_{X/\mathbb{C}})$. Thus, if $(9.7)_1$ is satisfied then we may unambiguously set

$$\langle \omega, \tau \rangle = \langle \widetilde{\omega}, \tau \rangle$$

and the equations

$$\langle \omega, \tau \rangle = 0 \ \text{ for } \ \omega \in H^0(\Omega^2_{X/\mathbb{C}})$$

are well-defined. If for $\partial/\partial\alpha \in \Omega^{1*}_{\mathbb{C}/\mathbb{Q}}$ we set

$$\langle \omega \otimes \partial/\partial\alpha, \tau \rangle = \{\langle \omega, \tau \rangle, \partial/\partial\alpha\}$$

where the curly brackets on the right-hand side is the pairing $\Omega^1_{\mathbb{C}/\mathbb{Q}} \otimes \Omega^{1*}_{\mathbb{C}/\mathbb{Q}} \to \mathbb{C}$, then the equations (9.7) are equivalent to $(9.7)_1$ together with

$$(9.7)_2 \qquad \langle \omega \otimes \partial/\partial\alpha, \tau \rangle = 0 \text{ for all } \omega \in H^0(\Omega^2_{X/\mathbb{C}}) \text{ and } \partial/\partial\alpha \in \Omega^{1*}_{\mathbb{C}/\mathbb{Q}}.$$

Finally, since ω is in the image of $H^0(\Omega^2_{X/\mathbb{Q}}) \to H^0(\Omega^2_{X/\mathbb{C}})$ we have

$$(9.10) \qquad \nabla(\omega \otimes \partial/\partial\alpha) = 0 \text{ in } H^1(\Omega^1_{X/\mathbb{C}}),$$

where ∇ is the arithmetic Gauss-Manin connection.

For X defined over a general field

$$\ker\{H^0(\Omega^2_{X/\mathbb{C}}) \otimes \Omega^{1*}_{\mathbb{C}/\mathbb{Q}} \xrightarrow{\nabla} H^1(\Omega^1_{X/\mathbb{C}})\}$$

is by [18] the image of

$$H^0(\Omega^2_{X/\mathbb{Q}}) \otimes \Omega^{1*}_{\mathbb{C}/\mathbb{Q}} \to H^0(\Omega^2_{X/\mathbb{C}}) \otimes \Omega^{1*}_{\mathbb{C}/\mathbb{Q}}.$$

The equations

$$(9.11)_1 \qquad \langle \varphi, \tau \rangle = 0 \ \text{ for all } \ \varphi \in H^0(\Omega^1_{X/\mathbb{C}}),$$

$$(9.11)_2 \qquad \langle \xi, \tau \rangle = 0 \ \text{ for all } \ \xi \in \ker\{H^0(\Omega^2_{X/\mathbb{C}}) \otimes \Omega^{1*}_{\mathbb{C}/\mathbb{Q}} \to H^1(\Omega^1_{X/\mathbb{C}})\}$$

are well-defined, and when X is defined over $\mathbb{Q}$ they reduce to $(9.7)_1$ and $(9.7)_2$. One may write them out in coordinates as was done in (9.9) above. This involves choosing an open covering $\{U_\alpha\}$ and writing

$$(\nabla\xi)_{\alpha\beta} = \delta\{\sigma_\alpha\}$$

for $\sigma = \{\sigma_\alpha\} \in C^1(\{U_\alpha\}, \Omega^1_{X/\mathbb{C}})$. Then the analogue of (9.9) will involve ξ and σ. We refer to the following discussion for a proof and to section 9.3 below, where explicit computations of this will be given for general surfaces in $\mathbb{P}^3$.

We now look again at Abel's differential equations from a different, more formal perspective and give proofs of the interpretations stated above. For this we consider the Cousin flasque resolution

$$0 \to \Omega^1_{X/\mathbb{Q}} \to \Omega^1_{\underline{\underline{\mathbb{C}}}(X)/\mathbb{Q}} \to \underline{\underline{T}}Z^1_1(X) \to \underline{\underline{T}}Z^2(X) \to 0.$$

Taking the hypercohomology spectral sequence of this and noting that all but the leftmost sheaf are flasque, we have an exact sequence

$$H^0(\underline{\underline{T}}Z^1_1(X)) \to H^0(\underline{\underline{T}}Z^2(X)) \to H^2(\Omega^1_{X/\mathbb{Q}}) \to 0,$$

or in other words

$$TZ^1_1(X) \to TZ^2(X) \to H^2(\Omega^1_{X/\mathbb{Q}}) \to 0.$$

We may thus say:

In terms of 9.6(a), Abel's differential equations for a surface are the map

$$TZ^2(X) \to H^2(\Omega^1_{X/\mathbb{Q}}).$$

From the exact sequence

$$0 \to \Omega^1_{\mathbb{C}/\mathbb{Q}} \otimes \mathcal{O}_X \to \Omega^1_{X/\mathbb{Q}} \to \Omega^1_{X/\mathbb{C}} \to 0$$

we have an exact sequence

$$0 \to \frac{\Omega^1_{\mathbb{C}/\mathbb{Q}} \otimes H^2(\mathcal{O}_X)}{\nabla_{X/\mathbb{Q}} H^1(\Omega^1_{X/\mathbb{C}})} \to H^2(\Omega^1_{X/\mathbb{Q}}) \to \ker(H^2(\Omega^1_{X/\mathbb{C}})$$

$$\xrightarrow{\nabla_{X/\mathbb{Q}}} \Omega^1_{\mathbb{C}/\mathbb{Q}} \otimes H^3(\mathcal{O}_X)) \to 0.$$

Here $\nabla_{X/\mathbb{Q}}$ is the arithmetic Gauss-Manin connection, and since X is a surface we have $H^3(\mathcal{O}_X) = 0$. For a regular surface,

$$H^2(\Omega^1_{X/\mathbb{Q}}) \cong \frac{\Omega^1_{\mathbb{C}/\mathbb{Q}} \otimes H^2(\mathcal{O}_X)}{\nabla_{X/\mathbb{Q}} H^1(\Omega^1_{X/\mathbb{C}})}.$$

When X is defined over $\bar{\mathbb{Q}}$, $\nabla_{X/\mathbb{Q}}$ is zero in the above expression and then

$$\begin{aligned} H^2\left(\Omega^1_{X/\mathbb{Q}}\right) &\cong \Omega^1_{\mathbb{C}/\mathbb{Q}} \otimes H^2(\mathcal{O}_X) \\ &\cong \operatorname{Hom}_{\mathbb{C}}\left(H^0(\Omega^2_{X/\mathbb{C}}), \Omega^1_{\mathbb{C}/\mathbb{Q}}\right). \end{aligned}$$

For such an X,

$$H^0\left(\Omega^2_{X(\bar{\mathbb{Q}})/\mathbb{Q}}\right) \otimes \mathbb{C} \twoheadrightarrow H^0\left(\Omega^2_{X/\mathbb{C}}\right),$$

so we may choose a basis of the holomorphic 2-forms that come from absolute Kähler differentials defined over $\bar{\mathbb{Q}}$.

We now explain how to associate to

$$\omega \in H^0\left(\Omega^2_{X(\bar{\mathbb{Q}})/\mathbb{Q}}\right)$$

an Abel's differential equation. If $\xi, \eta \in \bar{\mathbb{Q}}(X)$ give local uniformizing parameters around each point x_i,

$$x = (\xi, \eta) \in X$$

and if at x_i

$$\omega = h_i d\xi \wedge d\eta,$$

then for tangent vectors

$$v_i \in T_{x_i}X, \qquad v_i(\lambda_i, \mu_i),$$

the associated Abel's differential equation is

$$\sum_i h_i(x)(\mu_i d\xi_i - \lambda_i d\eta_i) = 0 \quad \text{in } \Omega^1_{\mathbb{C}/\mathbb{Q}},$$

where $\xi_i = \xi(x_i), \eta_i = \eta(x_i)$. More generally, if τ_i is a finite fourth quadrant Laurent tail at x_i with $\Omega^1_{X/\mathbb{Q}}\big|_{x_i}$ coefficients, then the associated Abel's differential equation is

$$\sum_i \mathrm{Res}_{x_i}(\omega\tau_i) = 0 \quad \text{in } \Omega^1_{\mathbb{C}/\mathbb{Q}}.$$

In the above, $\omega\tau_i$ is a Laurent tail with coefficients in the image of

$$\Omega^3_{X/\mathbb{Q}}\big|_{x_i} \to \Omega^2_{X/\mathbb{C}}\big|_{x_i} \otimes \Omega^1_{\mathbb{C}/\mathbb{Q}}$$

and we pull out the $\Omega^1_{\mathbb{C}/\mathbb{Q}}$ and then take the residue.

If X is regular and not defined over $\bar{\mathbb{Q}}$, but only over a field k (which we may take to be finitely generated over $\mathbb{Q}$), then letting

$$\mathrm{Der}_{k/\mathbb{Q}} = \left(\Omega^1_{k/\mathbb{Q}}\right)^*,$$

we get an Abel's differential equation associated to each element of

$$\ker\left(H^0(\Omega^2_{X(k)/k}) \otimes \mathrm{Der}_{k/\mathbb{Q}} \xrightarrow{\nabla_{X/\mathbb{Q}}} H^1\left(\Omega^1_{X(k)/k}\right)\right).$$

If ϕ belongs to this kernel, then on an open cover $\{U_\alpha\}$ of X, we may lift ϕ to

$$\widetilde{\phi}_\alpha \in H^0\left(U_\alpha, \Omega^2_{X(k)/\mathbb{Q}}/\Omega^2_{k/\mathbb{Q}}\right) \otimes \mathrm{Der}_{k/\mathbb{Q}}.$$

Now

$$\widetilde{\phi}_\alpha - \widetilde{\phi}_\beta \in H^0\left(U_\alpha \cap U_\beta, \Omega^1_{X(k)/k} \otimes \Omega^1_{k/\mathbb{Q}}\right) \otimes \mathrm{Der}_{k/\mathbb{Q}}.$$

If we let

$$\pi\left(\widetilde{\phi}_\alpha - \widetilde{\phi}_\beta\right) \in H^0\left(U_\alpha \cap U_\beta, \Omega^1_{X(k)/k}\right)$$

be obtained by contracting

$$\Omega^1_{k/\mathbb{Q}} \otimes \mathrm{Der}_{k/\mathbb{Q}} \to k,$$

then

$$[\pi(\widetilde{\phi}_\alpha - \widetilde{\phi}_\beta)] = 0 \quad \text{in } H^1\left(\Omega^1_{X(k)/k}\right).$$

Consequently we can find

$$\psi_\alpha \in H^0\left(U_\alpha, \Omega^1_{X(k)/k}\right)$$

with

$$\pi(\widetilde{\phi}_\alpha - \widetilde{\phi}_\beta) = \psi_\alpha - \psi_\beta \qquad \text{for all } \alpha, \beta.$$

With these preliminaries:

The Abel's differential equation associated to ϕ is, if $x_i \in U_{\alpha_i}$,

$$\sum_i \mathrm{Res}_{x_i}\left(\pi(\widetilde{\phi}_{\alpha_i}\tau_i) - \psi_{\alpha_i}\tau_i\right) = 0 \quad \text{in } \mathbb{C}.$$

Summary: We may give parallel formulations of Abel's differential equations for a curve and a surface as follows:

X a curve: For

$$\omega \in H^0\left(\Omega^1_{X/\mathbb{C}}\right), \quad \tau_i \in \mathcal{PP}_{X,x_i}$$

Abel's differential equations are

$$\sum_i \mathrm{Res}_{x_i}(\omega\tau_i) = 0.$$

X a surface: There are two sets of differential equations, one for 1-forms and one for 2-forms. We shall use the notation

$$LT_{X,x} = \{\text{finite } 4^{\text{th}} \text{ quadrant Laurent tails at } x\}.^1$$

Then for

$$\varphi \in H^0\left(\Omega^1_{X/\mathbb{C}}\right), \quad \tau_i \in LT_{X,x_i} \otimes \Omega^1_{X/\mathbb{Q},x_i}$$

Abel's differential equations for 1-forms are

$$\sum_i \mathrm{Res}_{x_i}(\varphi \wedge \tau_i) = 0.^2$$

Turning to 2-forms, first if X is defined over $\bar{\mathbb{Q}}$, then for

$$\omega \in H^0\left(\Omega^2_{X(\bar{\mathbb{Q}})/\mathbb{Q}}\right)$$

Abel's differential equations for 2-forms are

$$\sum_i \mathrm{Res}_{x_i}(\omega \wedge \tau_i) = 0 \quad \text{in } \Omega^2_{\mathbb{C}/\mathbb{Q}}.$$

If X is not defined over $\mathbb{Q}$, then there is a formulation using the arithmetic Gauss-Manin along the lines discussed above.

---

[1] For $\dim X = n$ we may denote by $LT_{X,x}$ the finite $(2^n)^{\text{th}}$ quadrant Laurent tails at x. Then for $n = 1$, $LT_{X,x} \cong \mathcal{PP}_{X,x}$, and in general

$$T_{\{x\}}Z^n(X) \cong LT_{X,x} \otimes \Omega^{n-1}_{X/\mathbb{Q},x}.$$

[2] Here, $\varphi \wedge \tau_i$ denotes the image in $LT_{X,i} \otimes \Omega^2_{X/\mathbb{C},x_i}$ using the natural map $\Omega^1_{X/\mathbb{C}} \otimes \Omega^1_{X/\mathbb{Q}} \to \Omega^2_{X/\mathbb{C}}$.

9.1.1 Appendix to section 9.1: Remarks on Mumford's argument and its extensions

Above we have discussed the result

(i) *If X is an algebraic surface with $h^{2,0}(X) \neq 0$, then a generic $z \in X^{(d)}$ does not move in a rational equivalence.*

As noted there, this result, which has a number of formulations and is due to Mumford, Roitman, Bloch-Srinivas, and Voisin, also follows from our infinitesimal theory. However, the term "generic" has somewhat different meanings in the different approaches, and here we want to briefly explain this.

The essential observation in Mumford's argument is the following: Let $\omega \in H^0(\Omega^2_{X/\mathbb{C}})$ and let

$$f : \mathbb{P}^r \to X^{(d)}$$

be a holomorphic mapping that has rank r at some point and where $f(\mathbb{P}^r)$ contains a general point of $X^{(d)}$. Then

$$\text{(ii)} \qquad r \leq d.$$

Here "general" has *geometric* meaning; we should have

$$\begin{cases} z = x_1 + \cdots + x_d \in f(\mathbb{P}^r), \\ x_i \neq x_j \text{ for } i \neq j \text{ and all } \omega(x_i) \neq 0. \end{cases}$$

Under these circumstances, by moving z slightly we may assume that $z = f(p)$, where $df(p)$ has rank r. Then $\operatorname{Tr}\omega$ is a symplectic form in $T^*_z X^{(d)}$, and since

$$f^*(\operatorname{Tr}\omega) = 0$$

it follows that

$$df(T_p\mathbb{P}^r) \subset T_z X^{(d)}$$

is a null plane for $\operatorname{Tr}\omega$ and (ii) follows.[3]

We note that X need only be a complex surface for this argument to work.

The estimate (ii) is asymptotically sharp, as follows by considering linear series on hypersurface sections of algebraic $K3$ surfaces.

The proof that a generic $z \in X^{(d)}$ does not move in a rational equivalence is based upon (ii) together with the following consideration: Let S be a smooth variety and

$$Z \subset X \times S$$

an algebraic cycle such that for a general point $s \in S$ the intersection

$$z_s = Z \cdot (X \times \{s\})$$

is a 0-cycle on X. Then we may define

$$\text{(iii)} \qquad \operatorname{Tr} Z : H^0(\Omega^2_{X/\mathbb{C}}) \to H^0(\Omega^2_{S/\mathbb{C}}).$$

---

[3]In his paper Mumford observes that the trace construction was used by Severi to study 0-cycles on a surface—only Severi drew the *wrong conclusion*. Mumford famously remarks that in this case the Italians' technique was superior to their vaunted intuition.

One definition is by using linearity in Z to reduce to the case where Z is an irreducible subvariety, which then gives a rational map

$$Z\colon\ S \to X^{(d)},$$

where $d = \deg z_s$. We may then set

$$\operatorname{Tr}_Z(\omega) = Z^*(\operatorname{Tr}\omega)$$

for $\omega \in H^0(\Omega^2_{X/\mathbb{C}})$. Another definition, to be used below, uses the Künneth components

$$[Z]_{p,q} \in H^p(X) \otimes H^q(S), \qquad p+q=4,$$

of the fundamental class of $[Z]$ of Z. Since $[Z]$ is a Hodge class, $[Z]_{2,2}$ has a component in

$$H^{0,2}(X) \otimes H^{2,0}(S) \simeq \operatorname{Hom}\left(H^0(\Omega^2_{X/\mathbb{C}}), H^0(\Omega^2_{S/\mathbb{C}})\right),$$

and this component is the map $\operatorname{Tr}_Z X$ in (ii).

Using this map we have

(iv) *Let Z_1, Z_2 be two algebraic cycles in $X \times S$ as above, and assume that for every point $s \in S$*

$$z_{1,s} \equiv_{\text{rat}} z_{2,s}.$$

Then, for $\omega \in H^0(\Omega^2_{X/\mathbb{C}})$,

$$\operatorname{Tr}_{Z_1}(\omega) = \operatorname{Tr}_{Z_2}(\omega).$$

For the proof, standard algebro-geometric arguments reduce us to the following situation: Replacing S by a finite covering if necessary, we have maps

$$\begin{cases} S \xrightarrow{f} X^{(k)}, \\ \mathbb{P}^1 \times S \xrightarrow{g} X^{(d+k)}, \end{cases}$$

such that

$$g(0,s) = z_{1,s} + f(s),$$

$$g(\infty,s) = z_{2,s} + f(s).$$

Then since

$$g^*(\operatorname{Tr}\omega) = \pi_S^*\varphi$$

for some $\varphi \in H^0(\Omega^2_{S/\mathbb{C}})$ we infer (iv).

Mumford then used (i) and (ii) to show that for the evident map

$$X^{(d)} \to CH^2(X)$$

we have

$$\dim(\operatorname{Im} X^{(d)}) \geqq d.$$

Roitman refined the argument to show

$$\dim(\operatorname{Im} X^{(d)}) = 2d.$$

This implies (i), where "generic" means "outside of a countable union of proper subvarieties." One may contrast this geometric meaning of generic with the more precise arithmetic/geometric meaning given by the infinitesimal theory above.

Remark: An easy observation is that

(v) *If* $h^0(K_X) \neq 0$, *then any dominant holomorphic mapping*

$$f\colon \mathbb{P}^1 \times S \to X^{(d)}$$

is constant in the $\mathbb{P}^1$*-factor.*

It is of interest to note that

(v) remains true if we only assume that

$$h^0(nK_X) \neq 0 \text{ for some } n \geqq 1.$$

Example: When X is an Enriques surface, then Bloch, Kas, and Lieberman proved that

$$CH^2(X) \cong \mathbb{Z}.$$

Here, $h^0(K_X) = 0$ but $h^0(2K_X) \neq 0$, so that (iv) applies. Thus for $z_1, z_2 \in X^{(d)}$ we have that there exist $w \in X^{(k)}$ and

$$f\colon \mathbb{P}^1 \to X^{(d+k)}$$

such that

$$f(0) = z_1 + w,$$

$$f(\infty) = z_2 + w,$$

but in general we cannot take $w = 0$. In other words, the equivalence

$$z_1 \equiv_{\text{rat}} z_2$$

cannot be taken to be given by a linear equivalence on an irreducible curve on X, but rather is given by a configuration of linear equivalences where cancellations are necessary.

For the proof we will show that $h^0(nK_X) \neq 0 \Rightarrow h^0(nK_{X^{(d)}}) \neq 0$. The case $n = 2$ will illustrate the idea. In local coordinates x, y on X let

$$\varphi = g(x, y)(dx \wedge dy)^2$$

be a quadratic differential. Then it is easy to see that the trace

$$\operatorname{Tr} \varphi = g(x_1, y_1)(dx_1 \wedge dy_1)^2 + \cdots + g(x_d, y_d)(dx_d \wedge dy_d)^2$$

is single-valued but has poles along the diagonals. On the other hand,

$$\operatorname{Tr} \varphi^{1/2} = g(x_1, y_1)^{1/2} dx_1 \wedge dy_1 + \cdots + g(x_d, y_d)^{1/2} dx_d \wedge dy_d$$

is holomorphic but not single-valued. But then

$$\Phi = \left(\frac{1}{d!}\right)^2 (\underbrace{\operatorname{Tr} \varphi^{1/2} \wedge \cdots \wedge \operatorname{Tr}}\ \varphi^{1/2})^2$$

$$= g(x_1, y_1) \cdots g(x_d, y_d)(dx_1 \wedge dy_1 \wedge \cdots \wedge dx_d \wedge dy_d)^2$$

is holomorphic and single-valued, and therefore gives a nonzero section of $2K_{X^{(d)}}$.

Turning to the proof of (iv), if f is not constant in the $\mathbb{P}^1$ factor we may assume that $\dim S = 2d - 1$. For Φ as above we have that for local reasons

$$f^*\Phi \neq 0$$

but for global reasons $h^0(2dK_{\mathbb{P}^1 \times S}) = 0$, which is a contradiction. □

9.2 ON THE INTEGRATION OF ABEL'S DIFFERENTIAL EQUATIONS

Above we have expressed in differential form the condition for infinitesimal rational motion of a 0-cycle on a curve or on a surface. Now we will discuss how one may "integrate" Abel's differential equations.

Definition: *By integration of Abel's differential equations we mean defining a Hodge-theoretic object $\mathcal{H}$ and constructing a map*

$$\psi : Z^n(X) \to \mathcal{H} \tag{9.12}$$

whose fibers are the rational equivalence classes of 0-cycles.

Two caveats: First, we shall always work modulo torsion. Second, what will be constructed is a sequence of such maps

$$\psi_i : \ker \psi_{i-1} \to \mathcal{H}_i, \qquad i = 0, 1, \ldots, n$$

(where ψ_{-1} is the trivial map) such that the intersection of the kernels is rational equivalence.

For $n = 1$ the construction of (9.12) is classical. Here we shall briefly review this, and then we shall introduce a Hodge-theoretic construction of a ψ in the case $n = 2$ for surfaces defined over $\mathbb{Q}$ whose fibers are exactly rational equivalence classes *provided that one assumes the conjecture of Bloch and Beilinson.* In [32] we shall give this construction in general.

We shall use the notation

$$Z^n(X)_i = \ker \psi_i.$$

In the case $n = 1$ of a smooth algebraic curve we have the classical maps

$$\psi_0 : Z^1(X) \to \mathbb{Z},$$
$$\psi_1 : Z^1(X)_0 \to J(X),$$

where $J(X)$ is the Jacobian variety of X. Here for $z = \sum_i n_i x_i$

$$\psi_0(z) = \sum_i n_i = \deg z,$$

and if $\psi_0(z) = 0$ so that $z = \partial\gamma$ for some 1-chain γ,

$$\psi_1(z) = \left(\int_\gamma \varphi_1, \ldots, \int_\gamma \varphi_g \right) \text{ mod periods,}$$

where $\varphi_1, \ldots, \varphi_g$ is a basis for $H^o(\Omega^1_{X/\mathbb{C}})$. We may think of ψ_0 as a bilinear map

$$Z^1(X) \otimes H^0(\Omega^0_{X/\mathbb{C}}) \xrightarrow{\psi_0} \mathbb{C}$$

given by

$$\psi_0(z, 1) = \int_z 1, \tag{9.13$_1$}$$

and of ψ_1 as a bilinear map

$$Z^1(X)_0 \otimes H^0(\Omega^1_{X/\mathbb{C}}) \xrightarrow{\psi_1} \mathbb{C} \text{ mod periods}$$

given by

$$\psi_1(z,\varphi) = \int_\gamma \varphi, \qquad \partial\gamma = z. \tag{9.13$_2$}$$

Now let X be an algebraic surface. Then we may define ψ_0 and ψ_1 exactly as in the curve case, where now ψ_1 maps to the Albanese variety. *The issue has always been how to define* ψ_2.

From $(9.13)_1$ and $(9.13)_2$ we want to define something like a bilinear map

$$Z^2(X)_1 \otimes H^0(\Omega^2_{X/\mathbb{C}}) \xrightarrow{\psi_2} \mathbb{C} \text{ mod periods}$$

given by

$$\psi_2(z,\omega) = \int_\Gamma \omega. \tag{9.13$_3$}$$

Here, Γ should be a 2-chain that can only be constructed if $\psi_0(z) = \psi_1(z) = 0$.

A first hint of how to proceed comes if we think of $H^0(\Omega^1_{X/\mathbb{C}})$ as a subspace of $T^*CH^n(X)$. When $n=1$ obviously $H^0(\Omega^1_{X/\mathbb{C}})$ is equal to $T^*CH^1(X)$. However, when $n=2$

$$T^*CH^2(X)/H^0(\Omega^1_{X/\mathbb{C}}) \cong \ker\left\{H^0(\Omega^2_{X/\mathbb{C}}) \otimes \Omega^{1*}_{\mathbb{C}/\mathbb{Q}} \xrightarrow{\nabla} H^1(\Omega^1_{X/\mathbb{C}})\right\},$$

where ∇ is the arithmetic Gauss-Manin connection. The right-hand side has to do with the obstruction to lifting differentials defined over $\mathbb{C}$ to differentials defined over $\mathbb{Q}$. Thus, $H^0(\Omega^2_{X/\mathbb{Q}})$ should enter the picture.

Suppose then that X is a regular surface defined over $\mathbb{Q}$, so that by base change

$$H^0(\Omega^2_{X(\mathbb{Q})/\mathbb{Q}}) \otimes \mathbb{C} \cong H^0(\Omega^2_{X(\mathbb{C})/\mathbb{C}}).$$

We want to define something like $(9.13)_2$ when $\omega \in H^0(\Omega^2_{X/\mathbb{Q}})$. Referring to the preceding section, for

$$\tau = \sum_i (x_i, \tau_i) \in TZ^2(X),$$

where $\tau_i \in T_{x_i}X$ and, for the sake of simplicity the x_i are assumed to be distinct, the condition that τ be tangent to infinitesimal rational motion is

$$\langle\omega,\tau\rangle =: \sum_i \langle\omega(x_i),\tau_i\rangle = 0 \qquad \text{in } \Omega^1_{\mathbb{C}/\mathbb{Q}} \tag{9.14}$$

for all ω. Thinking of $\Omega^1_{\mathbb{C}/\mathbb{Q}}$ geometrically as reflecting cotangent vectors to spreads, this relation suggests how to proceed to construct $(9.13)_2$.

Namely, suppose that $z \in Z^2\big(X(k)\big)_0$, set $\mathcal{X} = X \times S$ where $\mathbb{Q}(S) \cong k$, and let

$$\begin{array}{ccc} \mathcal{Z} & \subset & \mathcal{X} \\ \downarrow & & \downarrow{\scriptstyle \pi} \\ S & = & S \end{array}$$

be the spread of z. The form ω defines a form, denoted by $\widetilde{\omega}$, in $H^0(\Omega^2_{\mathcal{X}(\mathbb{Q})/\mathbb{Q}})$. As shown by the calculations in chapter 4, using the evaluation mappings in (9.14) above we may think of

$$\langle \omega(x_i), \tau_i \rangle \in \pi^* T_{s_0} S \subset T^*_{(x_i, s_0)} \mathcal{X};$$

that is, along $\mathcal{Z}$ *we may think of* $\widetilde{\omega}$ *as having one "vertical" and one "horizontal" foot*. This suggests that the 2-chain Γ in $(9.13)_2$ should be traced out by 1-chains γ_s in X parametrized by a 1-chain λ in S.

Thus, let $\lambda \in \Omega S$ be a closed loop in S. For each $s \in \lambda$ we have

$$z_s = \partial \gamma_s$$

for some 1-chain γ_s in X. Using that $H_1(X, \mathbb{Q}) = 0$ and that we are working modulo torsion, we will have

$$\gamma_1 = \gamma_0 + \partial \Lambda$$

for some 2-chain Λ. If we set

$$\Gamma = \bigcup_{s \in \lambda} \gamma_s + \Lambda$$

then

$$\partial \Gamma = \bigcup_{s \in \lambda} z_s,$$

and we define

$$I(z, \omega, \lambda) = \int_\Gamma \omega \qquad \text{mod periods.} \tag{9.15}$$

Again elementary considerations show that a different choice of paths γ_s changes (9.15) by a period $\int_\sigma \omega$, where

$$\sigma \in H_2(X, \mathbb{Z}).$$

Now recall that on a manifold M a *differential character* is given by a function on the loop space

$$\chi : \Omega M \to R$$

mapping to some abelian group R which is a quotient of $\mathbb{C}$ by a subgroup and which satisfies

$$\chi(\partial \Delta) \equiv \int_\Delta \varphi,$$

where Δ is a 2-chain and φ is a 2-form on M.

Proposition: *(i) If* $\lambda = \partial \Delta$ *then*

$$I(z, \omega, \lambda) = \int_\Delta \mathrm{Tr}_{\mathcal{Z}}(\widetilde{\omega}),$$

where $\mathrm{Tr}_{\mathcal{Z}}$ *is the trace of the generically finite map* $\mathcal{Z} \to S$. *Thus, (9.15) defines a differential character on* S.

(ii) $I(z, w, \lambda)$ *depends only on the k-rational equivalence class* $[z] \in CH^2(X(k))$.

Proof: (i) Let $\Sigma = \bigcup_{s\in\Delta} \gamma_s$ and $\mathcal{Z}_\Delta = \bigcup_{s\in\Delta} z_s$. Then

$$\partial\Sigma = \Gamma - \mathcal{Z}_\Delta$$

and the result follows from Stokes's theorem, noting that

$$\int_{\mathcal{Z}_\Delta} \widetilde{\omega} = \int_\Delta \mathrm{Tr}_{\mathcal{Z}}(\widetilde{\omega}).$$

(ii) Let z_t, for $t \in \mathbb{P}^1$, be a rational equivalence defined over k given by a cycle

$$\mathcal{Z} \subset \mathcal{X} \times \mathbb{P}^1.$$

Then

$$\frac{d}{dt} I(z_t, \omega, \lambda) = \int_\lambda \langle \omega, z_t' \rangle,$$

where z_t' is the tangent to the family of 0-cycles z_t and

$$\langle \omega, z_t' \rangle \in T_s^* S$$

when we use the evaluation maps $\Omega^1_{\mathbb{C}/\mathbb{Q}} \to T_s^* S$. Since $z_t' \in TZ^2(X)_{\mathrm{rat}}$, we have

$$\langle \omega, z_t' \rangle = 0$$

by the results in part (i) of this section (cf. (9.14) above). □

This argument shows how one may think of $I(z, \omega, \lambda)$ as "integrating" Abel's differential equations. Actually, as explained in [32] there is an extension of the construction of $I(z, \omega, \lambda)$ to all $\omega \in H^2(X)_{\mathrm{tr}}$—the transcendental part of $H^2(X, \mathbb{C})$—that does not show up infinitesimally. We will not go into this here, but rather shall note the following consequence which will also be established in the work referred to above:

Assuming the conjecture of Bloch-Beilinson, if $I(z, \omega, \lambda) = 0$ for all ω and λ, then

$$z \equiv_{\mathrm{rat}} 0$$

(modulo torsion).

This result has the following

Corollary: *If $h^{1,0}(S) = h^{2,0}(S) = 0$, then z is rationally equivalent to zero.*

We note that the conditions in the corollary are birationally invariant and thus depend only on the field k.

Addendum: For later reference we want to give the extension of the above construction to the case when X is defined over $\mathbb{Q}$ but may not be regular. Let $z \in Z^2\big(X(k)\big)$ be a 0-cycle of degree zero and satisfying

$$\psi_1(z) =: \mathrm{Alb}_X z = 0.$$

Since Alb (X) is defined over a finite extension field of $\mathbb{Q}$ and (i) is an algebraic condition, it follows that

$$\psi_1(z_s) = \mathrm{Alb}\, z_s = 0$$

for the spread $\{z_s\}_{s\in S}$. Put differently, given the spread picture

$$\begin{array}{ccc} \mathcal{Z} \subset \mathcal{X} = X \times S \\ \downarrow \quad \downarrow \\ S = S \end{array}$$

where we assume S is smooth and complete, we have that the maps induced from the Künneth component of the fundamental class of $\mathcal{Z}$

$$\mathcal{Z}_{*,0} : H_0(S, \mathbb{Q}) \to H_0(X, \mathbb{Q})$$

and

$$\mathcal{Z}_{x,1} : H_1(S, \mathbb{Q}) \to H_1(X, \mathbb{Q})$$

are both zero. Working modulo torsion, if λ is a closed curve on S, then

$$\mathcal{Z}(\lambda) = \{z_s\}_{s\in\lambda}$$

traces out a 1-cycle on X that is the boundary of a 2-chain Γ,

$$\mathcal{Z}(\lambda) = \partial\Gamma.$$

Since Γ is determined up to a 2-cycle on X, we may set as before

$$I(z, \omega, \lambda) = \int_\Gamma \omega \quad \text{mod periods.}$$

Here, $\omega \in H^0(\Omega^2_{X/\mathbb{C}})$ is all that is needed to make the definition. Again as before, if $\lambda = \partial\Delta$ then

$$I(z, \omega, \lambda) = \int_\Delta \mathrm{Tr}_{\mathcal{Z}}\omega.$$

To show that $I(z, \omega, \cdot)$ depends only on the k-rational equivalence class of z, we may argue as before using the description of the pairing

$$H^0(\Omega^2_{X/\mathbb{Q}}) \otimes TZ^2(X) \to \Omega^1_{\mathbb{C}/\mathbb{Q}}$$

satisfying

$$H^0(\Omega^2_{X/\mathbb{Q}}) \otimes TZ^2_{\mathrm{rat}}(X) \to 0.$$

For this argument we need that ω is the image in $H^0(\Omega^2_{X/\mathbb{C}})$ of some $\widetilde{\omega} \in H^0(\Omega^2_{X/\mathbb{Q}})$.

The relation between this construction and the preceding is that, when X is regular, we may take

$$\Gamma = \bigcup_{s\in\lambda} \gamma_s.$$

9.3 SURFACES IN $\mathbb{P}^3$

We shall prove the following

Proposition 9.16: *Let X be a general surface of degree $d \geqq 5$ in $\mathbb{P}^3$. Then*

$$TX \cap TZ^2_{\mathrm{rat}}(X) = 0.$$

This means that for *any* point $p \in X$ and nonzero $v \in T_pX$, the image of $\tau = (p, v)$ in $TZ^2(X)$ is *not* tangent to rational equivalence.

We intend Proposition 9.16 as an illustration of how the spread perspective suggests geometric approaches, rather than as a "state of the art" result. In fact, the following corollary due to Herb Clemens is well known and can be improved into a lower bound on the genus—see [10] and [11] for this result and for references to earlier work.

Corollary: *A generic surface of degree $d \geqq 5$ in $\mathbb{P}^3$ contains no rational curves.*[4]

Suppose the proposition is false. Then for a general X there is a point $p \in X$ such that

$$T_pX \cap TZ^2_{\mathrm{rat}}(X) \neq 0.$$

Let

$$\begin{array}{ccc} \mathcal{X} & \supset & \mathcal{P} \\ \downarrow & & \downarrow \\ S & = & S \end{array} \tag{9.17}$$

be the spread of (X, p) and set $k = \mathbb{Q}(S)$. Throughout the argument we shall allow ourselves to shrink S and to pass to finite coverings $S' \to S$; i.e., to pass to finite algebraic extensions k' of k. Thus we may assume that $\mathcal{X} = \{X_s\}_{s \in S}$ is a smooth family of smooth surfaces in $\mathbb{P}^3$ and $\mathcal{P} = \{p(s)\}_{s \in S}$ with $p(s) \in X$ is a cross-section of $\mathcal{X} \to S$. We denote by $(X, p) = \big(X_{s_0}, p(s_0)\big)$ the data corresponding to a generic point $s_0 \in S$. We then have

$$\Omega^1_{k/\mathbb{Q}} \cong T^*_{s_0}S \tag{9.18}$$

as k-vector spaces. By our assumption of genericity and since $d \geqq 5$, the Kodaira-Spencer mapping

$$\rho : T_{s_0}S \to H^1(\Theta_X)$$

is surjective. Using the identification (9.18) it follows that the extension class

$$e \in \mathrm{Hom}\left(\Omega^{1*}_{k/\mathbb{Q}}, H^1(\Theta_X)\right) \tag{9.19}$$

of

$$0 \to \Omega^1_{k/\mathbb{Q}} \otimes \mathcal{O}_{X(k)} \to \Omega^1_{X(k)/\mathbb{Q}} \to \Omega^1_{X(k)/k} \to 0$$

is *surjective*.

In first approximation, we may think of our situation in coordinates as follows: The surface X is defined by

$$F(x) = \sum_I a_I x^I = 0,$$

[4]The proposition is stronger than the corollary, since the former precludes moving in certain rational equivalences on symmetric products—that is, we can add to $v \in T_p(X)$ infinitesimal creation/annihilation notions.

where we are using multi-index notation, the a_I are independent transcendentals, and

$$p(a) = [p_0(a), p_1(a), p_2(a), p_3(a)]$$

where $a = (\ldots, a_I, \ldots)$. More precisely, k will be a finite algebraic extension of $\mathbb{Q}(\ldots, a_I, \ldots)$ and we are working over an open set in S where the a_I are local uniformizing parameters, so that expressions such as

$$\frac{\partial p_i(a)}{\partial a_I}$$

are defined. If $V^* = H^0(\mathcal{O}_{\mathbb{P}^3}(1))$, then $\Omega^1_{k/\mathbb{Q}}$ has basis da_I and

$$\Omega^{1*}_{k/\mathbb{Q}} \cong V^d,$$

where $V^d = \mathrm{Sym}^d V$. The idea is that the cohomology groups $H^0(\Omega^2_{X(k)/k})$ and $H^1(\Omega^1_{X(k)/k})$ and the arithmetic Gauss-Manin connection ∇ have polynomial descriptions, and that consequently the calculation of the pairing

$$\langle \xi, \tau \rangle,$$

where

$$\xi \in TZ^2_{\mathrm{rat}}\big(X(k)\big)^{\perp} \cong \ker\{H^0(\Omega^2_{X(k)/k} \otimes \Omega^{1*}_{k/\mathbb{Q}} \to H^1(\Omega^1_{X(k)/k})\}$$

can be reduced to algebra.

We shall now explain the main computational step in the proof. Set

$$I = \ker\{H^0(\Omega^2_{X(k)/k}) \otimes \Omega^{1*}_{k/\mathbb{Q}} \to H^1(\Omega^1_{X(k)/k})\}.$$

Let $F_i = F_{x_i}$ and let

$$J = \bigoplus_{k \geqq d-1} J^k = \mathrm{span}\{F_0, F_1, F_2, F_2\}$$

be the *Jacobi ideal*. Then it is well known that

$$H^0(\Omega^2_{X(k)/k}) \cong V^{d-4},$$

$$H^1(\Omega^1_{X(k)/k})_{\mathrm{prim}} \cong V^{2d-4}/J^{2d-4},$$

where $H^1(\Omega^1_{X(k)/k})_{\mathrm{prim}}$ is the primitive part of the cohomology,

$$H^1(\Theta_{X(k)}) \cong V^d/J^d$$

and by our choice of k

$$\Omega^{1*}_{k/\mathbb{Q}} \cong V^d.$$

Moreover the Kodaira-Spencer map ρ and arithmetic Gauss-Manin connection ∇ are given, using these identifications, by

$$\begin{cases} V^d \xrightarrow{\rho} V^d/J^d, \\ V^{d-4} \otimes V^d \xrightarrow{\nabla} V^{2d-4}/J^{2d-4}, \end{cases}$$

where the second map is just polynomial multiplication. We set

$$K = \ker\{V^{d-4} \otimes V^d \to V^{2d-4}/J^{2d-4}\}.$$

Then

$$K = \left\{ \xi = \sum_I H_I \otimes x^I : \sum_I H_I x^I = \sum_i R_i F_i \right\}, \tag{9.20}$$

where $H_I \in V^{d-4}$ and $R_i \in V^{d-3}$. Recalling that $p = p(\ldots, a_I, \ldots)$ is a cross-section of $\mathcal{X} \to S$ we define

$$W_\xi(p) = \big(W_{\xi,0}(p), W_{\xi,1}(p), W_{\xi,2}(p), W_{\xi,3}(p)\big),$$

where

$$W_{\xi,i}(p) = \sum_I H_I\big(p(a)\big)\frac{\partial p_i(a)}{\partial a_I} + R_i\big(p(a)\big). \tag{9.21}$$

Lemma 1: *$W_\xi \in T_pX$, and up to a nonzero scale factor*

$$\langle \xi, \tau \rangle = \begin{vmatrix} p_0 & p_1 & p_2 & p_3 \\ v_0 & v_1 & v_2 & v_3 \\ W_{\xi,0}(p) & W_{\xi,1}(p) & W_{\xi,2}(p) & W_{\xi,3}(p) \end{vmatrix}. \tag{9.22}$$

Here, T_pX is being identified with $V/k \cdot p$, so that $k \cdot p + T_pX$ is given by

$$\left\{ w = (w_0, w_1, w_2, w_3) : \sum_i F_i(p) w_i = 0 \right\},$$

and $\tau = (p, v) \in T_pX$ where $v = (v_0, v_1, v_2, v_3)$ in (9.22) above.

The proof of the proposition will follow from (9.22) together with

Lemma 2: *If $d \geqq 5$, then for every $p \in X$*

$$W : K \to T_pX$$

given by $\xi \to W_\xi(p)$, is surjective.

Proof of Lemma 1: We will compute in affine coordinates and then express the result in homogeneous coordinates. Thus, let $\mathcal{U} \subset \mathbb{P}^3$ be one of the standard affine open sets with coordinates x_1, x_2, x_3 obtained by setting one of the homogeneous coordinates, say x_0, equal to one. Denote by f, h the affine versions obtained from F, H by setting $x_0 = 1$. Then we have on X

$$f_1 dx_1 + f_2 dx_2 + f_3 dx_3 = -\sum_I f_{a_I} da_I = -\sum_I x^I da_I$$

since $f = \sum_I a_I x^I$. This gives the equality

$$f_1 dx_1 \wedge dx_3 + f_2 dx_2 \wedge dx_3 = -\sum_I x^I da_I \wedge dx_3$$

of absolute differentials on X. On the open set where $f_1 \neq 0$, $f_2 \neq 0$ this gives

$$\frac{h dx_2 \wedge dx_3}{f_1} = \frac{h dx_3 \wedge dx_1}{f_2} - \frac{\sum_I h x^I da_I \wedge dx_3}{f_1 f_2}, \tag{i}$$

where $\deg h \leqq d-4$. If we are considering ordinary differential forms—i.e., sections of $\Omega^2_{X(k)/k}$—then the right-hand term drops out and the equality expresses the fact that the *Poincaré residue* of $\Omega = (h dx_1 \wedge dx_2 \wedge dx_3)/f$ may be computed in a well-defined manner by writing

$$\Omega = \omega \wedge \frac{df}{f}$$

and mapping

$$\Omega \to \omega\big|_X .$$

However, as expressed by (i) the Poincaré residue operator is not well-defined for absolute differentials, and this is the central point of the computation.

Suppose now that

$$\xi = \sum_I h_I \otimes x^I \in I$$

so that we have

(ii)
$$\sum_I h_I x^I = \sum_i r_i f_i .$$

Then using

$$v_1 f_1(p) + v_2 f_2(p) + v_3 f_3(p) = 0$$

for $v = (v_1, v_2, v_3) \in T_p X$ we obtain from (i) and (ii) that

(iii)
$$\begin{aligned}
&\sum_I \frac{\partial}{\partial a_I} \rfloor \left(\frac{h_I(v_2 dp_3 - v_3 dp_2)}{f_1(p)} \right) - \sum_I \frac{\partial}{\partial a_I} \rfloor \left(\frac{h_I(v_3 dp_1 - v_1 dp_3)}{f_2(p)} \right) \\
&\qquad = - \sum_I \frac{h_I x^I v_3}{f_1(p) f_2(p)} \\
&\qquad = - \frac{r_1 f_1(p) v_3 - r_2 f_2(p) v_3 + r_3 f_1(p) v_1 + r_3 f_2(p) v_2}{f_1(p) f_2(p)} \\
&\qquad = \frac{r_3 v_2 - r_2 v_3}{f_1(p)} - \frac{r_1 v_3 - r_3 v_1}{f_2(p)} .
\end{aligned}$$

Expressing this in homogeneous coordinates we find that the expression

(iv)
$$\sum_I \frac{\partial}{\partial a_I} \rfloor H_I(p) \frac{\begin{vmatrix} p_0 & p_1 & p_2 & p_3 \\ v_0 & v_1 & v_2 & v_3 \\ dp_0 & dp_1 & dp_2 & dp_3 \end{vmatrix}}{F_1(p)} + \frac{\begin{vmatrix} p_0 & p_1 & p_2 & p_3 \\ v_0 & v_1 & v_2 & v_3 \\ R_0(p) & R_1(p) & R_2(p) & R_3(p) \end{vmatrix}}{F_1(p)},$$

which is defined and regular in the open set $U_1 = \{F_1 = 0\}$, is equal to the same expression with $F_1(p)$ replaced by $F_2(p)$ when we consider it in U_2.

Now we are essentially done. Set $\Omega_I = (h_I dx_1 \wedge dx_2 \wedge dx_3)/f$ and denote by $\omega_I \in H^o(\Omega^2_{X(k)/k})$ the usual Poincaré residue of Ω_I. Denote by

$$\widetilde{\omega}_{I,1} = h_I \frac{dx_2 \wedge dx_3}{f_1}$$

the above lifting of ω_I to an absolute differential in U_1, and similarly for U_2 and U_3. Our hypothesis is that

$$\text{(v)} \qquad \sum_I \rho\left(\frac{\partial}{\partial a_I}\right)\omega_I = 0 \text{ in } H^1(\Omega^1_{X(k)/k}).$$

The left-hand side of this equation is computed, relative to the covering U_i, by lifting $\omega_I | U_i$ to an absolute differential as above, and then taking in $U_1 \cap U_2$

$$\sum_I \frac{\partial}{\partial a_I} \rfloor (\widetilde{\omega}_{I,2} - \widetilde{\omega}_{I,2}).$$

By assumption this is a coboundary $\sigma_1 - \sigma_2$, and expression (iv) above gives an explicit expression for this coboundary. Over $U_1 \cap U_2$ we thus have

$$\text{(vi)} \qquad \sum_I \frac{\partial}{\partial a_I} \rfloor \widetilde{\omega}_{I,1} + \sigma_1 = \sum_I \frac{\partial}{\partial a_I} \rfloor \widetilde{\omega}_{I,2} + \sigma_2.$$

We now follow the prescription in section 6.1 for evaluating an absolute 2-form on $v \in T_p X$ to obtain an element of $\Omega^1_{k/\mathbb{Q}}$, given over U_1 by the expression in (iv) above.

To complete the proof of the lemma we differentiate $0 = F(p) = \sum_I a_I p^I$ to obtain

$$\frac{\partial F}{\partial a_I}(p) + \sum_j \frac{\partial p_j}{\partial a_I} F_j(p) = 0.$$

This gives

$$\sum_I H_I(p) \frac{\partial F}{\partial a_I}(p) + \sum H_I(p) \frac{\partial p_j}{\partial a_I} F_j(p) = 0.$$

Using $\sum_I H_I(x) x^I = \sum_j R_j(x) F_j(x)$ at $x = p$ gives

$$\sum_I H_I(p) p^I = \sum_j R_j(p) F_j(p)$$

$$\Rightarrow \sum_j \left(\sum_I H_I(p) \frac{\partial p_j}{\partial a_I} + R_j(p) \right) F_j(p) = 0.$$

The term in the parenthesis is just $W_{\xi,j}(p)$, which is also the jth component in the bottom row of the determinants in (iv). □

Proof of Lemma 2: Suppose that there exist c_j such that

$$\sum_j c_j W_{\xi,j}(p) = 0$$

for all $\xi \in K$. We will show that

$$\text{(9.23)} \qquad c_j = \lambda F_j(p)$$

for some λ. The proof will be done in two steps.

Step one: Let

$$K_0 = \ker\{V^{d-4} \otimes V^d \to V^{2d-4}\}.$$

We will show that

$$W_\xi(p) = 0 \text{ for all } \xi \in K_0 \Leftrightarrow \frac{\partial p_j}{\partial a_I} = \lambda_j p^I \tag{9.24}$$

for some $\lambda_0, \lambda_1, \lambda_2, \lambda_3$.

For this we have that K_0 is generated by relations of the form

$$x_i R \otimes x_j S - x_j R \otimes x_i S. \tag{9.25}$$

In fact, any relation of the form

$$x^I \otimes x^J - x^{I'} \otimes x^{J'},$$

where $|I| = |I'| = d-4$, $|J| = |J'| = d$ and $I + J = I' + J'$, is generated by relations (9.25) as they allow us to "trade" x_i and x_j across the tensor product.

We now define the sheaf $\mathcal{R}$ over $\mathbb{P}^3$ by

$$0 \to \mathcal{R} \to V^d \otimes \mathcal{O}_{\mathbb{P}^3} \to \mathcal{O}_{\mathbb{P}^3}(d) \to 0.$$

We claim that

$$H^o(\mathcal{R}(1)) \otimes \mathcal{O}_{\mathbb{P}^3} \to \mathcal{R}(1) \to 0 \tag{9.26}$$

is surjective. For this we first observe that

$$\text{image}\,\{V^{d-1} \otimes \Lambda^2 V \to V^d \otimes V\} \subseteq H^o(\mathcal{R}(1)).$$

Since we are dealing with vector bundles it is enough to check (9.26) pointwise. Since the entire question is $GL(V)$ equivariant, it is enough to check that, in homogeneous coordinates x_0, x_1, x_2, x_3 at $p = [1, 0, 0, 0]$,

$$\text{image}\,\{V^{d-1} \otimes \Lambda^2 V \to V^d \otimes V\}_p = \mathcal{R}(1)_p.$$

Now

$$\mathcal{R}(1)_p = \bigoplus \mathcal{O}_{\mathbb{P}^3}(1)_p$$

when the direct sum is over all monomials except x_0^d. Since the map $V^{d-1} \otimes \Lambda^2 V \to V^d \otimes V \to \mathcal{R}(1)_p$ is given by

$$x^I \otimes x_j \wedge x_k \to x^{I+j} \otimes \delta_{0k} - x^{I+k} \otimes \delta_{0j}$$

we see that

$$x^I \otimes x_j \wedge x_0 \to x^{I+j} \otimes 1$$

if $j \neq 0$. Thus we hit every monomial except x_0^d and this establishes (9.26).

We now observe that, for $d \geqq 5$,

$$H^o(\mathcal{R}(d-4)) \otimes \mathcal{O}_{\mathbb{P}^3} \to \mathcal{R}(d-4) \to 0 \tag{9.27}$$

is surjective. This follows from (9.26), since for $d \geqq 5$

$$H^o(\mathcal{O}_{\mathbb{P}^3}(d-5)) \otimes \mathcal{O}_{\mathbb{P}^3} \to \mathcal{O}_{\mathbb{P}^3}(d-5) \to 0$$

is surjective.

We will now prove (9.24). The composite map

$$K_0 = H^o(\mathcal{R}(d-4)) \to \mathcal{R}(d-4)_p \to V^d \otimes \mathcal{O}_{\mathbb{P}^3}(d-4)_p \tag{9.28}$$

is

$$\xi = \sum_I H_I \otimes x^I \to \sum_I x^I \otimes H_I(p)$$

where $\sum_I H_I(x)x^I = 0$. If

$$c = \sum_I (x^I)^* \otimes c_I \in (V^d)^*$$

annihilates the image of (9.28), so that

$$\sum_I H_I(p)c_I = 0 \text{ for all } \xi \in K_0, \tag{9.29}$$

then since by (9.27)

$$(\mathrm{Im}\, K_0)^\perp = \mathcal{R}(d-4)_p^\perp,$$

$$\mathcal{R}(d-4)_p \subset V^d \otimes \mathcal{O}_{\mathbb{P}^3}(d-4)_p,$$

has codimension 1, and

$$\sum_i (x^I)^* \otimes p^I \in V^{d*} \otimes \mathcal{O}_{\mathbb{P}^3}(d)_p$$

is a generator of $\mathcal{R}(d-4)_p^\perp$, we infer from (9.29) that

$$c_I = \lambda p^I$$

for some constant λ. Applying this to $c_I = \frac{\partial p_j}{\partial a_I}$ for each j gives (9.24).

Step two: Now suppose that

$$\sum_j \sum_I H_I(p) c_j \frac{\partial p_j}{\partial a_I} + \sum_j c_j R_j(p) = 0 \tag{9.30}$$

for all $\xi \in K$. Using only K_0 we have (9.24). Substituting in the left-hand side of (9.30) we obtain for $\lambda = \sum_j c_j \lambda_j$

$$\begin{aligned} 0 &= \lambda\Big(\sum_I H_I(p)p^I\Big) + \sum_j c_j R_j(p) \\ &= \lambda\Big(\sum_j R_j(p)F_j(p)\Big) + \sum_j c_j R_j(p) \\ &= \sum_j R_j(p)(\lambda F_j(p) + c_j). \end{aligned}$$

Since this holds for all R_j of degree $d-3$ we have

$$\lambda F_j(p) + c_j = 0, \qquad j = 0, 1, 2, 3$$

as desired. □

It remains to discuss the argument when there exists a pair (X, p) with

$$T_pX \cap TZ^2_{\text{rat}}(X) \neq 0$$

defined over a general field k; say k is a finite algebraic extension of $\mathbb{Q}(\alpha_1, \ldots, \alpha_N)$. The important fact is that, if X is general, then the Kodaira-Spencer map

$$\Omega^{1*}_{k/\mathbb{Q}} \to H^1(\Theta_{X(k)}) \cong V^d/J^d$$

should be surjective. With this being so the previous argument can be easily extended to cover the more general case.

Analysis of the proof shows that essentially the same argument gives

(9.31) *For $X \subset \mathbb{P}^3$ a general surface of degree $d \geqq 6$ and p, q any distinct points of X we have*

$$\big(T_p(X) + T_q(X)\big) \cap TZ^2_{\text{rat}}(X) = 0$$

Corollary: *A general surface of degree at least six contains no curve having a g^1_2.*

We suspect that the method can be extended to give a proof of the

Conjecture: *Given k there is $d(k)$ such that a general surface in $\mathbb{P}^3$ of degree $d \geqq d(k)$ contains no curve having a g^1_k.*

Discussion: What is needed to prove (9.31) is first of all to replace (9.27) by the surjectivity of

$$H^o\big(\mathcal{R}(d-4)\big) \otimes \mathcal{O}_{\mathbb{P}^3} \to \mathcal{R}(d-4)_p \oplus \mathcal{R}(d-4)_q,$$

and this follows from (9.26) together with the surjectivity of

$$H^o\big(\mathcal{O}_{\mathbb{P}^3}(d-5)\big) \otimes \mathcal{O}_{\mathbb{P}^3} \to \mathcal{O}_{\mathbb{P}^3}(d-5)_p \oplus \mathcal{O}_{\mathbb{P}^3}(d-5)_q$$

for $d \geqq 6$. This means that effectively in the proof of (9.24) we may treat p and q as acting independently—i.e., we may repeat the argument just below (9.23) where the $H_I(q) = 0$. This gives (9.24) for each of p and q. Then under (9.30) we may repeat the argument to conclude the result.

9.3.1 Appendix A to section 9.3: On a theorem of Voisin

In [31] the following result is established:[5]

(i) *Let $X \subset \mathbb{P}^3$ be a general surface of degree at least seven. Then no two distinct points of X are rationally equivalent.*

This result is similar to Proposition 9.16 above, and Voisin's proof has the similarities that infinitesimal methods are used and that the argument is reduced to polynomial algebra. But the result does not follow from our descriptions of $TZ^2(X)$ and $TCH^2(X)$. Rather the proof uses the holomorphic 2-forms on X in a fashion

[5] Cf. [24], where a related result is proved.

similar to that discussed in the appendix to section 6.1 above, but now X is allowed to vary.

In outline the argument goes as follows: We consider the situation

$$\begin{array}{ccc} \mathcal{X} & \supset & \mathcal{Z} \\ \downarrow & & \downarrow \\ S & = & S \end{array}$$

where $\mathcal{X} = \{X_s\}_{s\in S}$ is the family of smooth surfaces of degree d in $\mathbb{P}^3$ (passing here to a finite covering of the moduli space and restricting to a Zariski open set), and $\mathcal{Z} = \{Z_s\}_{s\in S}$ is a family of 0-cycles of degree zero on the X_s. The fundamental class

$$[\mathcal{Z}] \in H^2(\Omega^2_{\mathcal{X}/\mathbb{C}}).$$

Also, the Leray spectral sequence associated to the filtration

$$F^p\Omega^{p+q}_{\mathcal{X}/\mathbb{C}} = \mathrm{image}\{\Omega^p_{S/\mathbb{C}} \otimes \Omega^q_{\mathcal{X}/\mathbb{C}} \to \Omega^{p+q}_{\mathcal{X}/\mathbb{C}}\}$$

degenerates at E_2 (cf. [18]). Moreover, the terms

$$\begin{cases} [\mathcal{Z}]^{4,0} = 0 & \text{in } E_2^{4,0} \text{ since } \deg z_s = 0, \\ [\mathcal{Z}]^{3,1} = 0 & \text{in } E_2^{3,2} \text{ since } h^3(X_s) = 0. \end{cases}$$

The $E_2^{2,2}$-term is expressed in terms of the variation of Hodge structure associated to $\mathcal{X} \to S$ in a well-known manner

$$E_2^{2,2} = (H^2(\Omega^2_{S/\mathbb{C}} \otimes \mathcal{H}^2))_\nabla,$$

where

$$\begin{cases} \mathcal{H}^2 = \mathbb{R}^2_\pi \Omega^{\cdot}_{\mathcal{X}/S}, \\ \nabla = \text{Gauss-Manin connection}, \\ (\quad)_\nabla \text{is the cohomology computed from } \nabla \end{cases}$$

Passing to the quotient $\mathcal{H}^2 \to R^2_\pi \mathcal{O}_{\mathcal{X}}$, the class $[\mathcal{Z}]^{2,2}$ gives

(ii) $$\delta\mathcal{Z} \in H^0(\Omega^2_{S/\mathbb{C}} \otimes \mathcal{H}^{0,2})/\nabla H^0(\Omega^1_{S/\mathbb{C}} \otimes \mathcal{H}^{1,1}).$$

This $\delta\mathcal{Z}$ is the extension to variable X of the trace construction discussed above. In the language of S. Saito [20] it may be thought of as a *higher normal function.*

Next, using a variant of the argument of Bloch-Srinivas [17], Voisin infers that if for a general $s \in S$

$$z_s \equiv_{\mathrm{rat}} 0,$$

then modifying $\mathcal{Z}$ by a rational equivalence—which does not change $[\mathcal{Z}]$—we may assume that $\pi(\mathcal{Z})$ is supported in a proper subvariety of S. It follows at a general point

(iii) $$\delta\mathcal{Z}(s) = 0.$$

We now assume that for a general point $s \in S$ there are distinct points $p(s), q(s) \in X_s$ such that

$$p(s) \equiv_{\text{rat}} q(s).$$

Taking $z_s = p(s) - q(s)$ we are in the above situation. The final geometric step is to show by explicit computation that for a general s

$$\text{(iv)} \qquad \delta \mathcal{Z}(s) \neq 0$$

in contradiction to (iii). Shrinking S if necessary, $H^0(\Omega^2_{S/\mathbb{C}} \otimes \mathcal{H}^{0,2})$ is the space of sections of a vector bundle with fibers

$$\frac{\Lambda^2 T_s^* S \otimes H^2(\mathcal{O}_{X_s})}{\nabla\left(T_s^* S \otimes H^1\left(\Omega^1_{X_s/\mathbb{C}}\right)\right)} \cong \operatorname{Hom}\left(H^0(\Omega^2_{X_s/\mathbb{C}}), \Lambda^2 T_s^* S\right).$$

The right-hand side of this isomorphism has a polynomial description, and calculations very similar to those above may now be used to establish (iv). We refer to [31] for details.

At this juncture, very roughly speaking,

> *we may say that for arguments using 2-forms "generic" means "outside a countable union of proper subvarieties," while for those using $TZ^p(X)$ and $TCH^2(X)$ "generic" refers to the "transcendental independence of suitable coefficients of defining relations."*

9.4 EXAMPLE: $(\mathbb{P}^2, T)$

The one example where the nonclassical part of the Chow group is nontrivial and understood seems to be the case of 0-cycles on the relative variety $(\mathbb{P}^2, T)$, where T is a triangle. We will explain this result, which is due to Bloch and Suslin (cf. [7], [21], [26]), and discuss how it relates to our infinitesimal story. Before doing that we want to set a context for why one might be interested in $(\mathbb{P}^2, T)$.

In the early days of algebraic geometry it was recognized that the integral

$$\text{(9.32)} \qquad \int \frac{dx}{\sqrt{x^3 + ax + b}}$$

was fundamentally different from the trigonometric/logarithmic integrals that arise from integrals of rational differentials on rational curves. One might hope to gain some insight into (9.32) by degenerating a smooth cubic into one with an ordinary double point. The singular curve is a $\mathbb{P}^1$ with two points, say 0 and ∞, identified. Functions on the singular curve are given by rational functions f on $\mathbb{P}^1$ with $f(0) = f(\infty)$, and the limit of the integrand in (9.33) is a rational differential on $\mathbb{P}^1$ with logarithmic singularities at 0 and ∞. Setting $T_0 = \{0, \infty\}$ we are studying the relative curve $(\mathbb{P}^1, T_0)$ and the limit of (9.32) on the singular curve is easily understood.

Specifically, denoting by x a coordinate on $\mathbb{C}^* \subset \mathbb{P}^1$ we consider 0-cycles

$$z = \sum_i n_i x_i, \qquad x_i \in \mathbb{P}^1 \backslash T_0 \cong \mathbb{C}^*,$$

and the (mixed) Hodge-theoretic conditions that z be the divisor of a function $f \in \mathbb{C}(\mathbb{P}^1)$ with $f(0) = f(\infty)$ are

$$(9.33) \qquad \begin{aligned} &\text{(i)} \qquad \psi_0(z) = \textstyle\int_z 1 = 0, \\ &\text{(ii)} \qquad \psi_1(z) = \textstyle\int_\gamma \omega \equiv 0 \text{ mod periods,} \end{aligned}$$

where $\partial\gamma = z$ and $\omega = dx/x$. Conditions (i) and (ii) may of course be expressed in closed form as

$$\sum_i n_i = 0,$$

$$\prod_i x_i^{n_i} = 1.$$

We note that x is determined up to scaling $x \to \lambda x$, and $\prod_i x_i^{n_i}$ is independent of the scaling if (i) is satisfied. In summary we have isomorphisms

$$\begin{cases} \psi_0 : CH^1(X) \to \mathbb{Z}, \\ \psi_1 : CH^1(X)_0 \to \mathbb{C}^*, \end{cases}$$

where $CH^1(X)_0$ is the kernel of ψ_0.

Implicit in the above discussion is the following point: If for $a \in \mathbb{P}^1 \backslash T_0 \cong \mathbb{C}^*$ we denote by $(a) \in Z^1(\mathbb{P}^1, T_0)$ that point considered as a 0-cycle, then we may define a map

$$\mathbb{C}^* \to CH^1(\mathbb{P}^1, T)_0$$

by

$$a \to (a) - (1).$$

Then *this map is a group homomorphism.* For the proof we set

$$f = \frac{(x - ab)(x - 1)}{(x - a)(x - b)} \in \mathbb{C}(\mathbb{P}^1, T_0)$$

and observe that

$$(9.34) \qquad \begin{aligned} &\operatorname{div} f = (ab) - (a) - (b) + (1) \\ &\Rightarrow (ab) - (1) \equiv_{\text{rat}} (a) - (1) + (b) - (1). \end{aligned}$$

Actually, anticipating what will happen below for surfaces, in terms of equations we could degenerate the above cubic into what is in some ways the simplest singular cubic, namely a triangle T

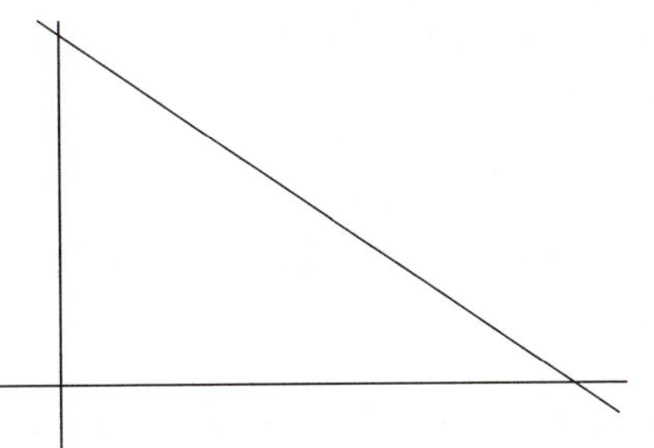

Each side of T is a $\mathbb{P}^1$ with two marked points, and the above discussion of $(\mathbb{P}^1, T_0)$ can be easily extended to the triangle case using the fact that the limit of the regular differential $dx/\sqrt{x^3+ax+b}$ on the smooth cubic induces on T a differential with logarithmic singularities and with opposite residues at the vertices.

When we turn to algebraic surfaces, say smooth surfaces $X \subset \mathbb{P}^3$, the first one for which $CH^2(X)$ is not understood occurs when $\deg X = 4$. By analogy with the curve case above, one might degenerate X into a tetrahedron X_0 and seek to understand $CH^2(X_0)$. Now the faces of a tetrahedron may be considered as relative surfaces biregularly equivalent to $(\mathbb{P}^2, T)$, and it turns out that one may also describe $CH^2(X_0)$ by understanding $CH^2(\mathbb{P}^2, T)$. The latter is what we shall now describe.

We identify $\mathbb{P}^2 \backslash T$ with $\mathbb{C}^* \times \mathbb{C}^*$ having coordinates x, y; these are unique up to scalings. The (mixed) Hodge theory of $(\mathbb{P}^2, T)$ has as basis the differential forms

$$\begin{cases} 1 \in F^0 H^0(\Omega^0_{\mathbb{P}^2}(\log T)), \\ \frac{dx}{x}, \frac{dy}{y} \in F^1 H^0\big(\Omega^1_{\mathbb{P}^2}(\log T)\big), \\ \frac{dx}{x} \wedge \frac{dy}{y} \in F^2 H^0\big(\Omega^2_{\mathbb{P}^2}(\log T)\big). \end{cases}$$

We may write 0-cycles $z \in Z^2(\mathbb{P}^2, T) = Z^2(\mathbb{P}^2 \backslash T)$ as

$$z = \sum_i n_i (x_i, y_i),$$

and define

$$\psi_0 : Z^2(\mathbb{P}^2, T) \to \mathbb{Z},$$

$$\psi_1 : Z^2(\mathbb{P}^2, T)_0 \to \mathbb{C} \oplus \mathbb{C} \text{ mod periods}$$

by

$$\psi_0(z) = \int_z 1 = \sum_i n_i$$

and

$$\psi_1(z) = \left(\int_\gamma \frac{dx}{x}, \int_\gamma \frac{dy}{y} \right)$$

where $\partial\gamma = z$. The periods are in $2\pi\sqrt{-1}\,\mathbb{Z} = \mathbb{Z}(1)$, and identifying $\mathbb{C}/\mathbb{Z}(1)$ with $\mathbb{C}^*$ we have as in the case of $(\mathbb{P}^1, T_0)$

$$\psi_1(z) = \left(\prod_i x_i^{n_i}, \prod_i y_i^{n_i} \right) \in \mathbb{C}^* \oplus \mathbb{C}^*.$$

Since $\sum_i n_i = 0$ this is independent of the scalings.

To complete the story, recall that in section 9.2 above we have discussed how one may define

$$\psi_2(z) = \int_\Gamma \frac{dx}{x} \wedge \frac{dy}{y} \text{ mod periods}$$

in case $\psi_0(z) = \psi_1(z) = 0$. We will turn to this below after we have identified $CH^2(X)_1$.

We set

$$z_{a,b} = (a, b) - (a, 1) - (1, b) + (1, 1) \in Z^2(\mathbb{P}^2, T)_1.$$

In the following picture of the projective plane the dotted lines all look like $(\mathbb{P}^1, T_0)$:

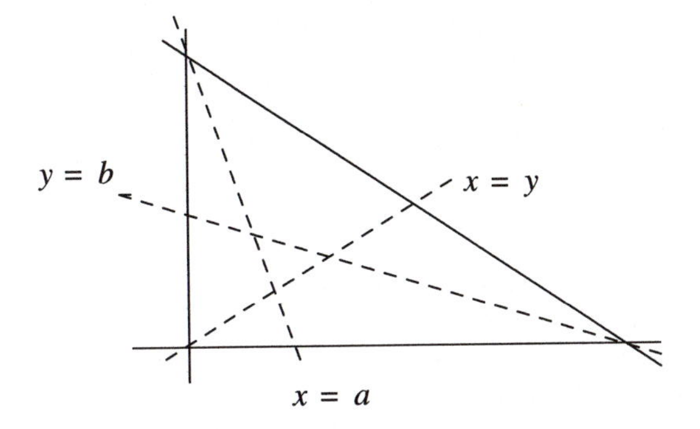

Thus, for example, by the argument that gives (9.34)

$$\begin{cases} z_{a_1 a_2, b} \equiv_{\text{rat}} z_{a_1,b} + z_{a_2,b} & \text{(using } y = b\text{)}, \\ z_{a, b_1 b_2} \equiv_{\text{rat}} z_{a,b_1} + z_{a,b_2} & \text{(using } x = a\text{)}, \\ z_{ab,ab} \equiv_{\text{rat}} 0 & \text{(using } x = y\text{)}. \end{cases}$$

Expanding the third relation using the first two gives

$$z_{a,b} + z_{b,a} \equiv_{\text{rat}} 0.$$

Letting $\equiv_{\text{rat}} \subset Z^2(\mathbb{P}^2, T)_1$ be the equivalence relation generated by divisors of rational functions on algebraic curves Y that have the same regular value on $Y \cap T$, the above shows that we have a well-defined map

$$Z^2(\mathbb{P}^2, T)_1 / \equiv_{\text{rat}} \to \mathbb{C}^* \otimes_{\mathbb{Z}} \mathbb{C}^*$$

given by

$$z_{a,b} \to a \otimes b - b \otimes a$$

and whose image lies in $\Lambda^2_{\mathbb{Z}} \mathbb{C}^*$.

We have now used all the lines that meet the triangle in two points (all the lines $x = \lambda y$ are equivalent after scaling). Consider now the line L given by

$$x + y = 1$$

and on L the function

$$f = \prod_i (x - a_i)^{n_i}.$$

Then if $\sum_i n_i = 0$ and

$$\prod_i a_i^{n_i} = \prod_i (1 - a_i)^{n_i} = 1$$

we have that $f \in \mathbb{C}(L, L \cap T)$ and thus

$$\sum_i n_i z_{a_i, 1-a_i} \equiv_{\text{rat}} 0.$$

This shows that the mapping

$$\Lambda^2_{\mathbb{Z}} \mathbb{C}^* \to CH^2(\mathbb{P}^2, T)$$

given by $a \wedge b \to z_{a,b}$ cannot be injective, and that in fact for *rather simple geometric reasons Steinberg relations necessarily enter.* We shall show that:

Proposition: *The mapping*

$$Z^2(\mathbb{P}^2, T)_1 \to K_2(\mathbb{C})$$

given by

$$\sum_i n_i(x_i, y_i) \to \prod_i \{x_i, y_i\}^{n_i}$$

induces an isomorphism

$$CH^2(\mathbb{P}^2, T)_1 \xrightarrow{\varphi} K_2(\mathbb{C}). \tag{9.35}$$

Proof: The proof proceeds in two steps. The proof of Totaro [9] was helpful in explaining the role of Suslin reciprocity here.

Step one: *The mapping φ in (9.35) is well-defined.*
Surprisingly, this is the harder step, since the second step will give an existence result. We have to show the following:

Let Y be a smooth algebraic curve and

$$Y \xrightarrow{F} \mathbb{P}^2$$

a regular mapping which maps generically one-to-one onto its image. Let

$$w = \sum_i n_i p_i, \qquad p_i \in Y$$

with $F(p_i) = (x_i, y_i)$, so that

$$F(w) = z.$$

Set $D = F^{-1}(T)$ and let $h \in \mathbb{C}(Y, D)$; i.e., $h \in \mathbb{C}(Y)$ and $h = 1$ on D. Then if

$$\operatorname{div} h = \sum_i n_i p_i$$

it follows from the *Suslin reciprocity law* that

$$\prod_i \{x_i, y_i\}^{n_i} = 1 \in K_2(\mathbb{C}). \tag{9.36}$$

We now sketch how this goes. For $p \in Y$ we define

$$\partial_p : K_m^M(\mathbb{C}(Y)) \to K_{m-1}^M(\mathbb{C})$$

by

$$\partial_p\{f_1, \dots, f_m\} = \prod_{i<j} \{T_p\{f_i, f_m\}, f_1(p), \dots, \hat{f_i}, \dots, \hat{f_j}, \dots, f_m(p)\}^{i+j-1}, \tag{9.37}$$

where we need a slightly more delicate definition if f_k has a zero or a pole at p for some $i \neq j$. Then Suslin's generalization of the Weil reciprocity law is

$$\prod_{p \in Y} \partial_p\{f_1, \dots, f_m\} = 1 \in K_{m-1}^M(\mathbb{C}).$$

Now suppose that $Y \to \mathbb{P}^2$ is given in affine coordinates by (f, g), where $f, g \in \mathbb{C}(Y)$. Then by (9.37) where $f_1 = f$, $f_2 = g$ and $f_3 = h$

$$\prod_i \{f(p_i), g(p_i)\}^{n_i} \underbrace{\prod_{\operatorname{div} f} \{g, h\} \prod_{\operatorname{div} g} \{h, f\}} = 1.$$

The last two terms over the bracket are equal to one since $\{1, a\} = 1$ in $K_2(\mathbb{C})$ and $h = 1$ on div $\cup$ div g.

Step two: *The mapping φ in (9.35) is injective.*

For the proof we consider the map

$$Z^1(\mathbb{P}^1 - \{0, 1, \infty\}) \xrightarrow{\text{St}} \Lambda^2_{\mathbb{Z}} \mathbb{C}^*$$

given by

(i) $$a \to a \wedge (1 - a).$$

In $Z^1(\mathbb{P}^1 - \{0, 1, \infty\})$ we consider the subgroup of all

$$\left\{ \sum_i n_i a_i \colon \sum_i n_i = 0, \prod_i a_i^{n_i} = \prod_i (1 - a)^{n_i} = 1 \right\},$$

which under (i) maps by

(ii) $$\sum_i n_i a_i \to \sum_i n_i a_i \wedge (1 - a_i).$$

It will suffice to show that

(iii) $$\text{Image (i)} = \text{Image (ii)}$$

Indeed, we have seen above that

$$\text{Image (i)}/ \equiv_{\text{rat}} \supseteq \text{Image (ii)}$$

and by definition

$$K_2(\mathbb{C}) = \Lambda^2_{\mathbb{Z}} \mathbb{C}^* / \text{image (i)},$$

from which our claim follows.

We set $B_2 = \ker \text{St}$ (this is the well-known *Bloch group*). One set of elements in B_2 is, for any distinct points $a_1, \dots, a_5 \in \mathbb{P}^1 - \{0, 1, \infty\}$,

(iv) $$\sum_\lambda (-1)^\lambda \text{CR}(a_1, \dots, \widehat{a_\lambda}, \dots, a_5),$$

where CR is closely related to the cross-ratio. (The fact that St(iv) $= 0$ is closely related to the famous 5-term functional equation for the dilogarithm.) In particular, for any distinct points $a, b \in \mathbb{P}^1 - \{0, 1, \infty\}$

$$(a) - (b) + \left(\frac{b(1-a)}{b-a}\right) - \left(\frac{b}{b-a}\right) + \left(\frac{b-1}{b-a}\right) \in B_2.$$

Now

$$\frac{a\left(\frac{b(1-a)}{b-a}\right) - \left(\frac{b-1}{b-a}\right)}{b\left(\frac{b}{b-a}\right)} = \frac{a(a-1)(b-1)}{b(b-a)}$$

and

$$\frac{(1-a)\left(1-\frac{b(1-a)}{b-a}\right)\left(1-\frac{b-1}{b-a}\right)}{(1-b)\left(1-\frac{b}{b-a}\right)}=\frac{(1-a)^2}{b-a}$$

For all $\alpha, \beta \in \mathbb{C}^*$ except for $\alpha = -\beta$ one may solve

$$\text{(v)} \qquad \begin{cases} \frac{a(1-a)(b-1)}{b(b-a)} = \alpha, \\ \frac{1-a)^2}{b-a} = \beta \end{cases}$$

for $a, b \in \mathbb{P}^1 - \{0, 1, \infty\}$. In the exceptional case $\alpha = -\beta$, equation (v) forces $a = b = 1$; in this case by a product of such relations one can obtain any $\alpha, \beta \in \mathbb{C}^*$.

Given a product

$$\prod_\nu (c_\nu \wedge (1-c_\nu))^{m_\nu} \in \text{Image St},$$

with

$$\prod_\nu c_\nu^{m_\nu} = \alpha, \prod_\nu (1-c_\nu)^{m_\nu} = \beta, \qquad \sum_\nu m_\nu = m$$

one may use an element of B_2 to change to the same element in Image St with $\alpha = \beta = 1$. One may then use an element of B_2 to change to the same element in Image St with $m \to m-1$. This establishes (iii) and completes the proof of the proposition. $\square$

Remark. Proof analysis shows that rational equivalence of 0-cycles in $\mathbb{P}^2 - T$ is generated by divisors of rational functions f or lines L such that $f = 1$ or $L \cap T$.

In general, for any algebraic surface X the relation of rational equivalence is generated by choosing a fixed very ample linear system $|L|$ and taking the divisors of rational functions on curves $Y \in |L|$.

Although we have not formally developed our infinitesimal theory for $TZ^p(X, Y)$ and $TCH^p(X, Y)$ in the relative case, it seems reasonable that this might be done. Making the identification

$$F^2H^0\left(\Omega^2_{\mathbb{P}^2/\mathbb{C}}(\log T)\right) \cong \mathbb{C}\omega,$$

where $\omega = (dx/x) \wedge (dy/y)$, one would then expect that

$$\text{(9.38)} \qquad TGr^2CH^2(\mathbb{P}^2, T) \cong \Omega^1_{\mathbb{C}/\mathbb{Q}}.$$

On the other hand, if we combine (9.35) with van der Kallen's theorem we obtain a potentially different way to make the identification (9.38). We will show that this does not happen. More precisely, let

$$z_t = \sum n_i \left(x_i(t), y_i(t)\right)$$

be an arc in $Z^1(\mathbb{P}^2, T)_1$ with tangent z' at $t = 0$. Then from the general theory we would expect that

$$\text{(9.39)} \qquad \langle \omega, z' \rangle = \sum_i n_i \left(\frac{x_i'}{x_i} \frac{dy_i}{y_i} - \frac{y_i'}{y_i} \frac{dx_i}{x_i} \right),$$

where $x_i' = dx^i(t)/dt$ at $t = 0$, $dx_i = d_{\mathbb{C}/\mathbb{Q}}\left(x_i(0)\right)$, and so on. We will show that

With $\langle \omega, z' \rangle$ defined by (9.39), we have

$$d\varphi\left(\frac{d}{dt}\right) = -\langle \omega, z' \rangle, \tag{9.40}$$

where the left-hand side is defined using van der Kallen's isomorphism.

Proof: This follows by writing, as in the proof of van der Kallen's theorem

$$\{x_0(1+\epsilon a), y_0(1+\epsilon b)\} = \{x_0, y_0\}\{x_0, 1+\epsilon b\}\{1+\epsilon a, 1+\epsilon b\}\{1+\epsilon a, y_0\}.$$

The first and third terms contribute zero when we take $d/d\epsilon$, and the second and fourth terms give

$$-b\frac{dx_0}{x_0} + a\frac{dy_0}{y_0},$$

where $d = d_{\mathbb{C}/\mathbb{Q}}$. □

A more conceptual argument follows from the commutative diagram below (6.37), noting that ω essentially gives the $d\log \wedge d\log$ mapping discussed in (6.37).

Finally, we want to discuss the integration of Abel's differential equations

$$\langle \omega, z' \rangle = 0.$$

Any $z \in Z^1(\mathbb{P}^2, T)_1$ may be written as

$$z = \sum_i z_{a_i, b_i}.$$

Considering an individual $z_{a,b}$, we suppose that $a, b \in k^*$, where $\operatorname{tr}\deg k = 1$, so that

$$k \cong \mathbb{Q}(S) \tag{9.41}$$

for some algebraic curve S. In section 9.2 above we have defined the differential character, defined for $\lambda \in H_1(S, \mathbb{Z})$ by

$$I(z_{a,b}, \omega, \lambda) = \int_\Gamma \omega \quad \text{mod periods.}$$

We may think of $I(z_{a,b}, \omega, \cdot)$ as landing in $H^1\big(S, \mathbb{C}/\mathbb{Z}(2)\big)$, and thus we have a map

$$\begin{array}{ccc} \mathbb{C}^* \times \mathbb{C}^* & \longrightarrow & H^1\big(S, \mathbb{Z}(2)\big) \\ \Cup & & \Cup \\ (a, b) & \longrightarrow & I(z_{a,b}, \omega, \cdot). \end{array}$$

Essentially because I is an invariant of rational equivalence, this induces a map

$$K_2(\mathbb{C}) \xrightarrow{I} H^1\big(S, \mathbb{C}/\mathbb{Z}(2)\big) \tag{9.42}$$

sending $\prod_i \{a_i, b_2\}$ to $\sum_i I(z_{a_i, b_i}, \omega, \cdot)$. On the other hand, there is a well-known map, the *regulator* (whose definition will be recalled below)

$$R : K_2\big(\mathbb{Q}(S)\big) \to H^1\big(S, \mathbb{C}/\mathbb{Z}(2)\big).$$

Proposition: *Using the identification (9.35) these two maps agree; i.e.*

$$R(\{a,b\}) = I(\{a,b\}). \tag{9.43}$$

Proof: For $S^o = S - D$ a Zariski open set in S we have the spread

$$\mathcal{Z} \subset (\mathbb{P}^2, T) \times S^o$$

of $z_{a,b}$; we write $\mathcal{Z} = \{z_{a(s),b(s)}\}_{s\in S}$. As S traces out a closed curve λ in S^o, $z_{a(s),b(s)}$ traces out $\mathcal{Z}(\lambda)$ in $\mathbb{P}^2 - T$. Since the homology class of $\mathcal{Z}(\lambda)$ is zero in $H_1(\mathbb{P}-T,\mathbb{Z})$, we may write

$$\mathcal{Z}(\lambda) = \partial\Gamma$$

and then

$$I(z_{a,b},\omega,\lambda) = \int_\Gamma \omega.$$

We have to rewrite the integral that will show it is equal to the usual definition of the regulator.

We write

$$z_{a(s),b(s)} = \big(a(s),b(s)\big) - \big(a(s),1\big) - \big(1,b(s)\big) + (1,1)$$

and may picture Γ as a surface with boundary

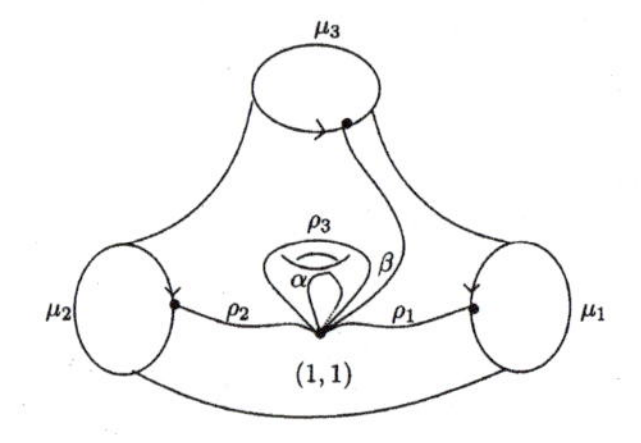

where

$$\begin{cases} \mu_1 = \{\text{the curve } \big(a(s),1\big)\} \\ \mu_2 = \{\text{the curve } \big(1,b(s)\big)\} \\ \mu_3 = \{\text{the curve } \big(a(s),b(s)\big)\} \end{cases}$$

and ρ_1,ρ_2,ρ_3 are the cuts indicated where $y=1$ along ρ_1 and $x=1$ along ρ_2. On the cut open surface we may write

$$\frac{dx}{x}\wedge\frac{dy}{y} = d\left(\log x\frac{dy}{y}\right).$$

Since $\mathcal{Z}(\lambda) = \mu_1+\mu_2+\mu_3$, by Stokes's theorem

$$\begin{aligned}\int_\Gamma \frac{dx}{x}\wedge\frac{dy}{y} &= \sum_i \int_{\mu_i} \log x\frac{dy}{y} - \int_{\rho_3}\frac{dy}{y}\int_{\mu_3}\frac{dx}{x} \\ &\quad + \int_{\rho_2}\frac{dy}{y}\int_{\mu_2}\frac{dx}{x} + \int_{\rho_1}\frac{dy}{y}\int_{\mu_1}\frac{dx}{x} \\ &\quad + \int_\alpha \log^+ x\frac{dy}{y} + \int_\beta \log^+ x\frac{dy}{y},\end{aligned}$$

where $\log^+ x$ is the jump in $\log x$ across the cuts α and β. The last two terms are in $\mathbb{Z}(2)$ and hence get factored out. In the first three terms, the one corresponding to μ_1 is zero and the one corresponding to μ_2 is in $\mathbb{Z}(2)$. Of the next three terms only

$$\int_{\rho_3} \frac{dy}{y} \int_{\mu_3} \frac{dx}{x} = \log y(s_0) \int_{\mu_3} \frac{dx}{x}$$

is nonzero, where $s_0 \in \lambda$ corresponds to our base point. Collecting the terms we have

$$\int_\Gamma \frac{dx}{x} \wedge \frac{dy}{y} = \int_\lambda \log x \frac{dy}{y} - \log y(s_0) \frac{dx}{x},$$

which is the usual expression for the regulator. □

For X a complete curve, we define $K_2(X)$ to be the subgroup of $K_2(\mathbb{C}(X))$ given by the kernel of the tame symbol maps T_x for all $x \in X$. Then the regulator defines a map

$$K_2(X) \xrightarrow{R} H^1(X, \mathbb{C}/\mathbb{Z}(2)).$$

A well-known conjecture is that (modulo torsion)

$$\ker R = K_2(\mathbb{C})$$

(cf. [37] for references and a discussion of image R).

Chapter Ten

Speculations and Questions

10.1 DEFINITIONAL ISSUES

There are three main questions:

(10.1) *Can one define $TZ^p(X)$ in general?*

The technical issue that arises in trying straightforwardly to extend the definitions given in the text for $p = n, 1$ concerns cycles that are linear combinations of irreducible subvarieties

$$Z = \sum_i n_i Z_i,$$

where some Z_i may not be the support of a locally Cohen-Maculay scheme. Similar issues are not unfamiliar—for example, in duality theory—and we can see no geometric reason why the question (10.1) should not have an affirmative answer.

The second main question is:

(10.2) *For $p = n, 1$ can one define $TZ^p(X)$ axiomatically?*

This issue has been raised several times in the text. Specifically, in section 6.1 we have in a significant example given a set of axioms that define $\equiv_{1^{\mathrm{st}}}$ geometrically and agree with the general definition. And again in sections 7.3, 7.4 we have illustrated and discussed the question (10.2).

Finally, an obvious question is

(10.3) *Can one define the Bloch-Gersten-Quillen sequence $\mathcal{G}_k$ on infinitesimal neighborhoods X_k so that*

(10.4) $$\ker\{\mathcal{G}_1 \to \mathcal{G}_0\} \cong \underline{\underline{T}}\mathcal{G}_0.$$

Here, X_k is $X \times \mathrm{Spec}(\mathbb{C}[\epsilon]/\epsilon^{k+1})$. Using the case $k = 1$ and $\dim X = 2$ for illustration, we are asking for something like

(10.5) $$0 \to \mathcal{K}_2\left(\mathcal{O}_{X_1}\right) \to \mathbb{C}(X_1)^* \to \bigoplus_{Y_1} \mathbb{C}(Y_1)^* \to \bigoplus_{x_1} \underline{\underline{\mathbb{Z}}}_{x_1} \to 0\,,$$

which gives a flasque "Cousin-type" resolution of $\mathcal{K}_2(\mathcal{O}_{X_1})$. There are obvious difficulties in defining the last two terms in (10.5). For example,

$$\bigoplus_{x_1} \underline{\underline{\mathbb{Z}}}_{x_1}$$

is supposed to be something like "the sum of skyscraper sheaves supported on equivalence classes of irreducible codimension-2 subschemes of X_1 that meet X_0

properly." The equivalence relation should be that two ideals $\mathcal{I}_1, \mathcal{I}_2$ in $\mathcal{O}_{X_1}$ are equivalent if they define the "same" irreducible subvariety of X_1. Simple examples based on the failure of the Nullstellesatz for nonreduced schemes show that, at least to us, there are significant difficulties in directly defining (10.4). In fact, the definition of $\underline{\underline{T}}Z^2(X)$ in the text may be thought of as giving the left-hand term in a possible exact sequence

$$0 \to \underline{\underline{T}}Z^2(X) \to \bigoplus_{x_1} \underline{\underline{\mathbb{Z}}}_{x_1} \to \bigoplus_{x} \underline{\underline{\mathbb{Z}}}_{x} \to 0. \tag{10.6}$$

Also the definition of $\underline{\underline{T}}Z_1^1(X)$ may be thought of as giving the first term in a possible exact sequence

$$0 \to \underline{\underline{T}}Z_1^1(X) \to \bigoplus_{Y_1} \underline{\underline{\mathbb{C}}}(Y)^* \to \bigoplus_{Y} \underline{\underline{\mathbb{C}}}(Y)^* \to 0, \tag{10.7}$$

and (10.6), (10.7) might fit in an exact sequence

$$0 \to \underline{\underline{T}}\mathcal{G}_0 \to \mathcal{G}_1 \to \mathcal{G}_0 \to 0$$

explaining in this case what would be meant by (10.4).

10.2 OBSTRUCTEDNESS ISSUES

We have defined, for $p = \dim X$ and $p = 1$, the tangent sheaf $\underline{\underline{T}}Z^p(X)$ and tangent space $TZ^p(X) = H^0\big(\underline{\underline{T}}Z^p(X)\big)$ to the space of codimension p algebraic cycles on a smooth variety X, and for the purpose of this discussion we shall assume that these definitions have been extended to all codimensions. We may think of a tangent vector as a first order variation of a cycle, and the *obstructedness question* asks when this may be successively extended to a formal infinite order variation of the cycle. The *convergence question* asks when the tangent vector is tangent to a geometric arc in the space of algebraic cycles. There are essentially four (not mutually exclusive) possibilities:

(i) $TZ^p(X)$ *may be obstructed.* This means that there exist X and $\tau \in TZ^p(X)$ such that, thinking of τ as a map

$$\operatorname{Spec}\left(\mathbb{C}[\epsilon]/\epsilon^2\right) \to Z^p(X), \tag{10.8}$$

this map cannot be lifted to a map

$$\operatorname{Spec}\left(\mathbb{C}[\epsilon]/\epsilon^{k+1}\right) \to Z^p(X)$$

for some $k \geqq 2$.

Remark: If this happens, then one expects that it already occurs for $k = 2$. The question of the equivalence on maps $\operatorname{Spec}(\mathbb{C}[\epsilon]/\epsilon^{k+1}) \to \operatorname{Hilb}^2(X)$ to define the "same" cycle has been raised in the preceding section.

(ii) $TZ^p(X)$ *is formally unobstructed.* This means that any tangent vector (10.1) may be lifted to a map

$$\lim_{k} \left(\operatorname{Spec}\mathbb{C}[\epsilon]/\epsilon^{k+1}\right) \to Z^p(X).$$

Remark: Let $Y \subset X$ be a smooth subvariety of codimension p and $\nu \in H^0(N_{Y/X})$ a normal vector field. It is well known that ν may be obstructed as an element of $T_Y\text{Hilb}^p(X)$; i.e., $\text{Hilb}^p(X)$ may be nonreduced at Y. However, as will be seen below, for $p = 1$ (and trivially for $p = n$) when we consider Y as a cycle in $Z^p(X)$ and ν as a tangent vector $\tau(\nu) \in TZ^p(X)$, the obstruction to lifting ν will disappear.

(iii) *$TZ^p(X)$ is formally unobstructed, but there exist $\tau \in TZ^p(X)$ that are not tangent to geometric arcs in $Z^p(X)$.*

One may ask similar questions at the sheaf level. The results proved in chapter 8 imply that:

(10.9) *For a smooth algebraic surface, every tangent vector in the stalks of the tangent sheaves*

$$\underline{\underline{T}}Z^1(X), \underline{\underline{T}}Z^2(X), \underline{\underline{T}}Z^1_1(X)$$

is tangent to a geometric arc in $Z^1(X)$, $Z^2(X)$, $Z^1_1(X)$ respectively.

Returning to the general discussion we have the fourth possibility:

(iv) *Every $\tau \in TZ^p(X)$ is tangent to a geometric arc in $Z^p(X)$.*

As to what is known about (i)–(iv), as noted in the introduction Ting Fai Ng has shown that for X smooth of any dimension

(10.10) *Every $\tau \in TZ^1(X)$ is tangent to a geometric arc in $Z^1(X)$.*

The geometric idea behind this result is quite simple and elegant. We illustrate it by explaining how, given a normal vector ν to a smooth curve Y on a smooth surface X, one may eliminate the first obstruction to lifting the corresponding tangent vector $\tau(\nu) \in TZ^1(X)$. It is well known that the obstruction $\mathcal{O}(\nu)$ to lifting ν viewed as a map

$$\text{Spec}(\mathbb{C}[\epsilon]/\epsilon^2) \to \text{Hilb}^1(X)$$

to $\text{Spec}(\mathbb{C}[\epsilon]/\epsilon^3)$ is in $H^1(N_{Y/X})$. Choose an ample divisor D on Y so that $H^1(N_{Y/X}(D)) = 0$. Then in terms of a Čech covering the image in $C^1(N_{Y/X}(D))$ of the obstruction cocycle $\mathcal{O}(\nu)$ may be written as a coboundary. What does this mean geometrically?

Suppose that $D = W \cap Y$, where $W \subset X$ is a sufficiently ample smooth curve. Then extending ν to be zero along W gives an element

$$\widetilde{\nu} \in H^0(N_{Y \cup W/X})$$

as depicted by

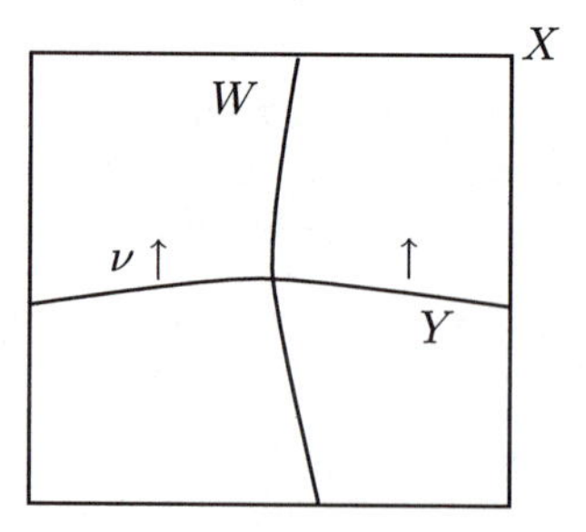

Here $N_{Y\cup W/X}$ is the normal sheaf $\mathrm{Hom}_{\mathcal{O}_X}(\mathcal{I}_{Y\cup W}/\mathcal{I}^2_{Y\cup W)}, \mathcal{O}_X)$ to the subscheme $Y \cup W$ of X. If $H^1(\mathcal{O}_X(Y+W)) = 0$, then Ng shows that the obstruction to lifting $\widetilde{\nu}$ to a map

$$\mathrm{Spec}(\mathbb{C}[\epsilon]/\epsilon^3) \to \mathrm{Hilb}^1(X),$$

which maps $\epsilon = 0$ to $Y \cup W$ vanishes. Thus writing

$$Y = (Y+W) - W$$

we may consider $\widetilde{\nu}$ as giving $\tau(\widetilde{\nu}) \in TZ^1(X)$, which may then be lifted to second order by moving $Y + W$ to second order and leaving $-W$ constant.

We may illustrate this locally around the point of intersection with coordinates x, y, where $Y = \{y = 0\}$ and $\nu = \partial/\partial y$. Then the first order deformation is locally the divisor of

$$f_1(x, y, t) = y - t.$$

Writing the obstruction as a coboundary with a first order pole at the origin we find that the second order deformation is given by taking the divisor of

$$\begin{aligned} f_2(x, y, t) &= y - t - \frac{t^2}{x} \\ &= \frac{1}{x}(xy - xt - t^2). \end{aligned}$$

Thus, $Y + W$ deforms into $xy - xt - t^2 = 0$, which is something like the picture

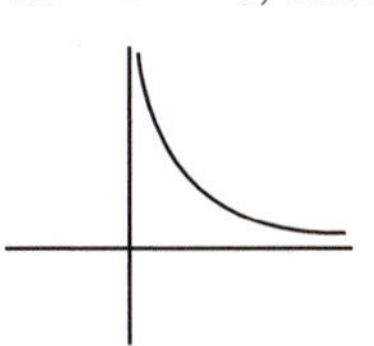

In contrast to the codimension 1 case, we have

(10.11) *For $p \geqq 2$, there exist X and tangent vectors $\tau \in TZ^p(X)$ that are not tangent to a geometric arc.*

The reason for this is as follows: Let X be a smooth threefold. The formal tangent space to the Chow group of codimension-2 cycles is given by

$$TCH^2(X) \cong H^2(\Omega^1_{X/\mathbb{Q}}). \tag{10.12}$$

The differential $\delta\psi_1$ of the Abel-Jacobi mapping

$$CH^2(X)_0 \xrightarrow{\psi_1} J^2(X)$$

is given, using the identification (10.5), by

$$H^2(\Omega^1_{X/\mathbb{Q}}) \xrightarrow{\delta\psi_1} H^2(\Omega^1_{X/\mathbb{C}}).$$

Thus

$$\mathrm{Image}\ \delta\psi_1 = \ker\{H^2(\Omega^1_{X/\mathbb{C}}) \xrightarrow{\nabla} H^3(\mathcal{O}_X) \otimes \Omega^1_{\mathbb{C}/\mathbb{Q}}\},$$

where ∇ is the arithmetic Gauss-Manin connection. In particular, if $H^3(\mathcal{O}_X) = 0$ or if X is defined over $\mathbb{Q}$, then

$$\text{Image } \delta\psi_1 = H^2(\Omega^1_{X/\mathbb{C}}).$$

Since

$$T(\text{Image } \psi_1) \subseteq H^2(\Omega^1_{X/\mathbb{C}})$$

corresponds to a *sub-Hodge structure*—i.e. it plus its conjugate is spanned by a $\mathbb{Q}$-lattice in $H^3(X, \mathbb{R})$—we know that in general Image $\delta\psi_1$ is too big. Since

$$TCH^2(X) \cong TZ^2(X)/TZ^2_{\text{rat}}(X)$$

we infer that we may have geometric tangent vectors in $TZ^2(X)$ that are not tangent to any geometric arc in the space of codimension 2 cycles.

Thus of the possibilities listed above, only (i)–(iii) can actually occur. Our guess is that

(ii) and (iii) above are the possibilities that actually occur for $p \geqq 2$.

10.3 NULL CURVES

Next we want to mention another curious phenomenon that arises in the infinitesimal theory of Chow varieties, this being what we call null curves. For simplicity, taking X to be a regular algebraic surface defined over $\mathbb{Q}$, a *null curve* is given by a picture

$$Y \subset B \times X, \tag{10.13}$$

where Y and B are algebraic curves with $Y \to B$ a branched covering such that the induced mapping

$$J(Y) \to CH^2(X)$$

is nonconstant but has differential that vanishes identically. (The terminology null-curve is introduced by analogy with the theory of relativity.) From the discussion in chapter 4 we see that

The diagram (10.13) defines a null-curve if it is defined over $\mathbb{Q}$.

The phenomenon of null-curves has, in our view, the following explanation: At each value of t, $z'(t)$ lies in the image of the tangent space to rational equivalences on X. However, this tangent vector to rational equivalences does not necessarily arise from a geometric family of rational equivalences. We expect that there is no obstruction to an infinite order *formal* lift of the tangent vector, but there is one going from a formal lift to a geometric family.

We will show elsewhere that, as a consequence of the Bloch-Beilinson conjectures, this is the only way that null-curves can arise.

10.4 ARITHMETIC AND GEOMETRIC ESTIMATES

One motivation for defining $TZ^p(X)$ is to have the possibility of using iterative methods to construct algebro-geometric objects—e.g., rational equivalences. Were $Z^p(X)$ a classical space—a manifold or a variety—one could hope to integrate Abel's differential equations by analytic methods. However, as just pointed out, this is not the case. Now solving a differential equation is an iterative geometric process, and as $Z^p(X)$ one might seek to devise iterative geometric/arithmetic processes that would in the limit produce algebro-geometric objects. Convergence of an iterative process requires estimates, and in closing we wish to offer some observations/speculations as to what form these might take.

For illustrative purposes we consider a regular algebraic surface X defined over $\mathbb{Q}$. Then the Bloch-Beilinson conjecture has the following implication:

(10.14) *Let $p, q \in X(\bar{\mathbb{Q}})$. Then there is an integer $d_{p,q}$ and a rational map*

$$f_{p,q} : \mathbb{P}^1 \to X^{(d_{p,q})}$$

such that

$$f_{p,q}(0) = p + z,$$
$$f_{p,q}(\infty) = q + z.$$

In fact, we may take $f_{p,q}$ to be defined over $\bar{\mathbb{Q}}$. Let $\delta_{p,q}$ be the degree of $f_{p,q}$ relative to some projective embedding of X—that is, $\delta_{p,q}$ is the degree of the curve traced out on X by the family of 0-cycles $f_{p,q}(t)$. We observe that

It is not possible to bound $d_{p,q}$ and $\delta_{p,q}$ for all $p, q \in X(\bar{\mathbb{Q}})$.

The reason is that if such bounds exist then by standard reasoning using Hilbert schemes and/or Chow varieties we will be able to infer that

$$p \equiv_{\text{rat}} q$$

for all $p, q \in X(\mathbb{C})$, which is not the case if $p_g(X) \neq 0$.

Let $H(p, q)$ be some measure of the arithmetic complexity of $p, q \in X(\bar{\mathbb{Q}})$. For example, relative to a projective embedding

$$X \to \mathbb{P}^N$$

defined over a finite extension of $\mathbb{Q}$, we may think of the coordinates of p and q as algebraic numbers. The arithmetic complexity of an algebraic number can then be measured by the heights of the coefficients of this equation. More generally, one may define the height of subvarieties defined over $\bar{\mathbb{Q}}$ (cf. [38]). Let $D(p, q)$ be some measure of the geometric size of (10.14); e.g., we might take

$$D(p, q) = d_{p,q} + \delta_{p,q}.$$

Then one might suspect that

$$H(p, q) \to \infty \Rightarrow D(p, q) \to \infty;$$

maybe there is even a bound

$$H(p, q) \leqq cD(p, q) \tag{10.14}$$

for some constant $c > 0$.

In other words, one may imagine that for a not necessarily regular algebraic surface defined over $\bar{\mathbb{Q}}$

(10.16) *If Bloch-Beilinson is true, then for a general* $z \in Z^2(X(\bar{\mathbb{Q}}))$ *satisfying*

$$\begin{cases} \deg z = 0, \\ \mathrm{Alb}\, z = 0, \end{cases}$$

the geometric size of any rational equivalence

$$z \equiv_{\mathrm{rat}} 0$$

is bounded from below by the arithmetic complexity of z.

In the one case where one knows Bloch-Beilinson, namely, the relative variety $(\mathbb{P}^2, T)$ discussed in section 9.4 above, we shall give heuristic reasoning in support of the converse to (10.16), namely,

(10.17) *For* $a, b \in \mathbb{Q}^*$, *there is a rational equivalence*

$$z_{a,b} \equiv_{\mathrm{rat}} 0$$

whose geometric size is bounded from above by the arithmetic complexity of a, b, *expressed as a computable function of the heights of* a, b.

The proof to be given also gives what we feel is strong evidence also for a lower bound in this case.

Proof: It is a result of Garland [40] that $K_2(\bar{\mathbb{Q}}) = 0$, i.e. in $\bar{\mathbb{Q}}^* \otimes_{\mathbb{Z}} \bar{\mathbb{Q}}^*$ we may write

$$a \otimes b = \prod_{\nu=1}^{N_{a,b}} c_\nu \otimes (1 - c_\nu), \tag{10.18}$$

where $c_\nu \in \bar{\mathbb{Q}}^*$. We claim that we may bound $N_{a,b}$ and the heights of the c_ν by the heights of a, b. One way to see this is as follows.

First, if $a, b \in \mathbb{Q}^*$ the theorem of Bass and Tate [35] gives that for some positive integer N

$$\{a, b\}^N = 1 \text{ in } K_2(\mathbb{Q}). \tag{10.19}$$

The proof is by writing $b = m/n$ where m and n are relative prime integers, and then using bilinearity and skew-symmetry of the Steinberg symbols to reduce to the case where $b = p$ is a prime. That $\{a, p\}$ is torsion then follows by an argument using Fermat's theorem. The number and arithmetic complexity of the Steinberg relations that are introduced, as well as the order N of torsion in (10.19), are bounded by the arithmetic complexity of a, b.

Next, an elementary argument shows that by adjoining roots of unity we may show that

$$\{a, b\} = 1 \text{ in } K_2(\bar{\mathbb{Q}}).$$

Again, the number and arithmetic complexity of the Steinberg relations that are introduced are bounded by the heights of a, b.

Next, as is shown in the proof of proposition (9.35), we have in $\Lambda^2_{\mathbb{Z}}\bar{\mathbb{Q}}^*$

$$\prod_\nu c_\nu \otimes (1 - c_\nu) \equiv \prod_i e_i \otimes (1 - e_i)^{n_i} \text{ modulo Im(St)} \tag{10.20}$$

where

$$\begin{cases} \sum_i n_i = 0, \\ \prod_i e_i^{n_i} = \prod_i (1-e_i)^{n_i} = 1. \end{cases}$$

Moreover, analysis of the proof of (9.35) shows that again the arithmetic complexities of the elements in $Z^1(\mathbb{P}^1 - \{0, 1, \infty\})$ that are introduced in the map

$$Z^1(\mathbb{P}^1 - \{0, 1, \infty\}) \xrightarrow{\text{St}} \Lambda^2_{\mathbb{Z}}\bar{\mathbb{Q}}^*$$

to obtain the congruence in (10.20) are bounded by a computable function of the heights of the c_ν.

Finally, referring to the explicit construction of a rational equivalence

$$\sum_i n_i z_{e_i, 1-e_i} \equiv_{\text{rat}} 0 \tag{10.21}$$

given in the proof of (9.26), we see that the *geometric size* of that particular rational equivalence $\equiv_{\text{rat}}$ is bounded by the *arithmetic complexity* of the right-hand side of (10.20). □

Remark: Presumably analysis of the proof of Garland's theorem would enable one to extend (10.17) to the case where a, b are algebraic numbers.

Returning to the case of an algebraic surface one may ask

(10.22) *For an algebraic cycle z satisfying (10.17), is there an iterative scheme*

$$z_i \to z_{i+1}, \qquad z_0 = z,$$

where $z_i \in Z^1(\bar{\mathbb{Q}})$ and

$$\begin{cases} z_i \equiv_{\text{rat}} z_{i+1}, \\ h(z_{i+1}) < h(z_i). \end{cases}$$

Such an iterative scheme is in fact implicit in the analysis of $(\mathbb{P}^2, T)$ discussed above. Evidently such a scheme would establish the Bloch-Beilinson conjecture for 0-cycles on a general algebraic surface defined over $\bar{\mathbb{Q}}$.

The absence of derivations of $\bar{\mathbb{Q}}$ means that the usual infinitesimal/geometric methods (differential equations) break down, at least on the face of it. Referring to (8.40), for a tangent vector

$$\tau = \sum_i (x_i, \tau_i)$$

where $x_i \in X(\bar{\mathbb{Q}})$ (assumed for simplicity to be distinct) and any $\tau_i \in T_{x_i}(X(\mathbb{C}))$ satisfying

$$\langle \varphi, \tau \rangle = 0, \qquad \varphi \in H^0\left(\Omega^1_{X/\mathbb{C}}\right)$$

we have

$$\tau \in TZ^2_{\text{rat}}(X).$$

Clearly, to "point" in the direction of an actual rational equivalence we must in general have

$$\tau_i \in T_{x_i}(X(\bar{\mathbb{Q}})).$$

Intuitively, one might like to choose τ_i to point in the direction of decreasing the height of x_i. At the moment any such scheme would seem to require completely new ideas.

In concluding we would like to formulate a problem that on the one hand is a sort of infinitesimal analogue of (10.22), and on the other hand is an interesting and perhaps double problem in arithmetic algebraic geometry. For this we assume that X is defined over a number field k and let

$$\tau \in T_z Z^2(X(k))$$

be a tangent to a 0-cycle z both defined over k and satisfying

$$\langle \varphi, \tau \rangle = 0 \qquad \text{for all } \varphi \in H^0\left(\Omega^1_{X(k)/k}\right).$$

Then we have shown that there are infinitesimal rational equivalences

$$\xi \in TZ^1_1(X(K))$$

defined over a finite extension field K of k and which map to τ under the rational mapping

$$TZ^1_1(X(K)) \to TZ^2(X(k)).$$

The problem is

(10.23) *Can one choose ξ so that the height of ξ is bounded by the height of τ?*

Essentially, constructing ξ amounts to writing a cocycle as a coboundary. We are asking if this can be done in a manner that controls heights.

Bibliography

[1] D. Mumford, Rational equivalence of 0-cycles on surfaces, *J. Math. Kyoto Univ.* **9** (1968), 195–204.

[2] A. A. Roitman, Rational equivalence of zero-dimensional cycles (Russian), *Mat. Zametki* **28**(1) (1980), 85–90, 169.

[3] A. A. Roitman, The torsion of the group of 0-cycles modulo rational equivalence, *Ann. of Math.* **111**(2) (1980), 553–569.

[4] S. Bloch, Lectures on algebraic cycles, *Duke Univ. Math. Ser.* **IV** (1980). Duke University Press.

[5] S. Bloch, K_2 and algebraic cycles, *Ann. of Math.* **99**(2) (1974), 349–379.

[6] J. Stienstra, On the formal completion of the Chow group $\mathrm{CH}^2(X)$ for a smooth projective surface in characteristic 0, *Nederl. Akad. Wetensch. Indag. Math.* **45**(3) (1983), 361–382.

[7] A. A. Suslin, Reciprocity laws and the stable rank of rings of polynomials (Russian), *Izv. Akad. Nauk SSSR Ser. Mat.* **43**(6) (1979), 1394–1429.

[8] U. Jannsen, Motivic sheaves and filtrations on Chow groups. Motives (Seattle, WA, 1991), 245–302, *Proc. Sympos. Pure Math.*, Part 1, **55** (1994), Amer. Math. Soc., Providence, RI.

[9] B. Totaro, Milnor K-theory is the simplest part of algebraic K-theory, *K-Theory* **6**(2) (1992), 177–189.

[10] L. Ein, Subvarieties of generic complete intersections, *Invent. Math.* **94**(1) (1988), 163–169.

[11] G. Xu, Subvarieties of general hypersurfaces in projective space, *J. Differential Geom.* **39**(1) (1994), 139–172.

[12] W. van der Kallen, Le K_2 des nombres duaux (French), *C. R. Acad. Sci. Paris Sér. A-B* **273** (1971), A1204–A1207.

[13] R. Harthshorne, Local cohomology, A seminar given by A. Grothendieck, Harvard University, Fall, 1961, *Lect. Notes in Math.* **20**, Springer-Verlag, Berlin-New York, 1967.

[14] R. Harthshorne, Residues and duality, Lecture notes of a seminar on the work of A. Grothendieck, given at Harvard 1963/64. With an appendix by P. Deligne, *Lect. Notes in Math.* **20**, Springer-Verlag, Berlin-New York, 1966.

[15] J. Lipman, Residues and traces of differential forms via Hochschild homology, *Contemp. Math.* **61**, A.M.S., Providence, RI, 1987.

[16] D. Quillen, Higher algebraic K-theory, I., *Algebraic K-theory, I: Higher K-theories (Proc. Conf., Battelle Mem. Inst., Seattle, WA, 1972)*, pp. 85–147. *Lect. Notes in Math.* **341**, Springer, Berlin, 1973.

[17] S. Bloch and V. Srinivas, Remarks on correspondences and algebraic cycles, *Amer. J. Math.* **105**(5) (1983), 1235–1253.

[18] H. Esnault and K. Paranjape, Remarks on absolute de Rham and absolute Hodge cycles (English. English, French summary), *C. R. Acad. Sci. Paris Sér. I Math.* **319**(1) (1994), 67–72.

[19] B. Angéniol and M. Lejeune-Jalabert, *Calcul différentiel et classes caractéristiques en géométrie algébrique, travaux en cours* [*Works in Progress*] **38**, Hermann, Paris, 1989.

[20] S. Saito, Motives, algebraic cycles and Hodge theory, in *The Arithmetic and Geometry of Algebraic Cycles (Banff, AB, 1998)*, 235–253, *CRM Proc. Lecture Notes* **24**, A.M.S., Providence, RI, 2000.

[21] A. Suslin, Algebraic K-theory of fields, *Proc. ICM* (1986), 222.

[22] D. Ramakrishnan, Regulators, algebraic cycles and values of L-functions, *Contemp. Math.* **83** (1989), 183–310.

[23] M. Green, Algebraic cycles and Hodge theory, Lecture notes (Banff).

[24] M. Green and P. Griffiths, An interesting 0-cycle, to appear in *Duke Math. J.*

[25] U. Jannsen, Equivalence relations on algebraic cycles, *The Arithmetic and Geometry of Algebraic Cycles (Banff, AB, 1998)*, 225–260, *NATO Sci. Ser. C Math. Phys. Sci.* **548**, Kluwer Acad. Publ., Dordrecht, 2000.

[26] S. Bloch, Algebraic cycles and higher K-theory, *Adv. Math.* **61**(3) (1986), 267–304.

[27] S. Bloch, On the tangent space to Quillen K-theory, *Lect. Notes in Math.* **341** (1973), Springer-Verlag.

[28] U. Jannsen, Motivic sheaves and filtration on Chow groups, *Proc. Sympos. Pure Math.* **55** (1994), A.M.S., Providence, RI, 245–302.

[29] J. P. Murre, On a conjectural filtration on the Chow groups at an algebraic variety, I., *Indag. Math.* **4** (1993), 177–188.

[30] R. L. Bryant, S. S. Chern, R. B. Gardner, H. L. Goldschmidt, P. A. Griffiths, Exterior differential systems, *M.S.R.I. Publ.* **18** (1991), Springer-Verlag, New York.

[31] C. Voisin, Variations de structure de Hodge et zéro-cycles sur les surfaces générales, *Math. Annalen* **299** (1994), 77–103.

[32] M. Green and P. Griffiths, Hodge theoretic invariants of algebraic cycles, *Internat. Math. Res. Notices* **9** (2003), 477–510.

[33] P. Griffiths, Variations on a theorem of Abel, *Invent. Math.* **35** (1976), 321–390.

[34] D. Eisenbud, Commutative algebra. With a view toward algebraic geometry, *Grad. Texts in Math.* **150** (1995), Springer-Verlag, New York.

[35] J. Milnor, Introduction to algebraic K-theory, *Ann. of Math. Studies* **72** (1971), Princeton University Press, Princeton, NJ; University of Tokyo Press, Tokyo.

[36] M. Green and P. Griffiths, Abel's differential equations, *Houston J. Math.* **28** (2002), 329–351.

[37] M. Green and P. Griffiths, The regulator map for a general curve, *Contemp. Math.* **312** (2002), 117–127.

[38] C. Soulé, Hermitian vector bundles on arithmetic varieties, *Proc. Symp. Pure Math.* **62.1** (1997), 383.

[39] Ting Fei Ng, Princeton University PhD thesis, in preperation.

[40] H. Garland, A finiteness theorem for K_2 of a number field, *Ann. of Math.* **94** (1971), 534–548.

[41] Stienstra, On K_2 and K_3 of truncated polynomial rings, Algebraic K-theory, Evanston, IL, 1980.

[42] B. Lawson, Spaces of Algebraic Cycles—Levels of Holomorphic Approximation, *Proc. ICM Zurich.*

[43] J.-P. Serre, Géométrie algébrique et géométrie analytique, *Ann. Inst. Fourier* 6 (1956), 1–42.

Index